Fossil Collecting in the Mid-Atlantic States

Fossil Collecting in

the Mid-Atlantic States

With Localities, Collecting Tips,
and Illustrations of More than
450 Fossil Specimens

Jasper Burns

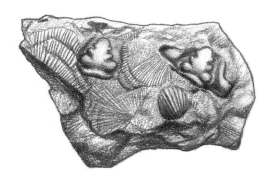

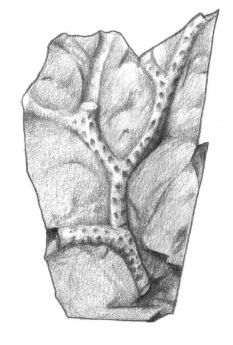

The Johns Hopkins University Press
Baltimore and London

A ROBERT G. MERRICK EDITION

Copyright © 1991 The Johns Hopkins University Press
All rights reserved
Printed in the United States of America on acid-free paper

04 03 02 01 00 99 98 97 96 95   6 5 4 3 2

The Johns Hopkins University Press
2715 North Charles Street
Baltimore, Maryland 21218-4319
The Johns Hopkins Press Ltd., London

Library of Congress Cataloging-in-Publication Data

Burns, Jasper.
    Fossil collecting in the Mid-Atlantic states : with
localities, collecting tips, and illustrations of more than
450 fossil specimens by Jasper Burns.
      p.   cm.
    Includes bibliographical references and index.
    ISBN 0-8018-4121-6.—ISBN 0-8018-4145-3 (pbk.)
    1. Fossils—Middle Atlantic States—Collection and
preservation—Guide-books.   I. Title.
QE718.B87   1991
560.9'75—dc20   90-45388 CIP

A catalog record for this book is available from the British Library.

To my parents,
James Richard and Jaquelin Caskie Burns

Contents

Preface ix

I. Understanding Fossils 1

What Is a Fossil? 3 / Setting the Stage: Geologic Time 4 / Knowing the Actors: The Many Names of Fossils 5 / How They Got There: A Brief Geologic History of the Area 8 / The Paleozoic Era 10 / The Mesozoic and the Cenozoic Eras 12 / Getting to Know the Sedimentary Rocks 14

II. How to Collect Fossils 21

How to Start 23 / Identifying What You Find 24 / Sizing Up a Road Cut 25 / Combing the Ancient Beach 27 / What You Will Need in the Field 28 / Respect Your Road Cut 31 / Flies in the Ointment: Hazards in the Field 32 / Back Home Again 33 / Where Do I Put Them? 36

III. A Portfolio of Fossil-Collecting Localities 39

1. Near Effinger, Rockbridge County, Virginia 44
2. Near Lusters Gate, Montgomery County, Virginia 46
3. Near Perry, Hardy County, West Virginia 48
4. At Swatara Gap, Lebanon County, Pennsylvania 51
5. Vicinity of Germany Valley Overlook, Pendleton County, West Virginia 53
6. Near Lexington, Rockbridge County, Virginia 56
7. Near Warm Springs, Bath County, Virginia 59
8. Near Cumberland, Allegany County, Maryland 60
9. Near Waiteville, Monroe County, West Virginia 62
10. Near Romney, Hampshire County, West Virginia 64
11. Along the Lost River near Wardensville, Hardy County, West Virginia 67
12. Along Thorn Creek near Franklin, Pendleton County, West Virginia 70
13. Near Fulks Run, Rockingham County, Virginia 73
14. On Bullpasture Mountain, Highland County, Virginia 74
15. Vicinity of Smoke Hole, Pendleton County, West Virginia 76
16. Near Monterey, Highland County, Virginia 81
17. Near the Lost River, Hardy County, West Virginia 84
18. Near Capon Lake, Hampshire County, West Virginia 88
19. East of Franklin, Pendleton County, West Virginia 91
20. Along the Cacapon River, near Yellow Spring, Hampshire County, West Virginia 93

21. Near Fort Frederick State Park, Washington County, Maryland   96
22. Gore, Frederick County, Virginia   98
23. East of Wardensville, Hardy County, West Virginia   101
24. Near Baker, Hardy County, West Virginia   104
25. In the Town of Lost City, Hardy County, West Virginia   107
26. South of Moorefield, Hardy County, West Virginia   110
27. North of Rio, Hampshire County, West Virginia   112
28. Near Gainesboro, Frederick County, Virginia   115
29. Near Hedgesville, Berkeley County, West Virginia   117
30. In Massanutten Mountain, Shenandoah County, Virginia   119
31. Near Edinburg, Shenandoah County, Virginia   120
32. Near Danville, Allegany County, Maryland   121
33. Near Flintstone, Allegany County, Maryland   122
34. Between Keenan and Gap Mills, Monroe County, West Virginia   124
35. In Locust Creek near Hillsboro, Pocahontas County, West Virginia   127
36. East of Uniontown, Fayette County, Pennsylvania   129
37. Along Scenic Route 150 Northwest of Marlinton, Pocahontas County, West Virginia   132
38. Across the Ohio River from Ambridge, Beaver County, Pennsylvania   135
39. Across the Ohio River from Sewickley, Allegheny County, Pennsylvania   138
40. Along the G. C. and P. Road near Wheeling, Ohio County, West Virginia   139
41. Near Bethany, Brooke County, West Virginia   142
42. Near Delaware City, New Castle County, Delaware   144
43. Along the shores of the Potomac River, Stafford and King George Counties, Virginia   146
44. Along the Calvert Cliffs on the Western Shore of the Chesapeake Bay, Calvert County, Maryland   150
45. Along the Shores of the Potomac River, Westmoreland County, Virginia   155
46. South Bank of the York River near the Mouth of Indian Field Creek, York County, Virginia   159

IV. Major Fossil Groups   165

Appendix: Geological Surveys in the Mid-Atlantic Region   181
Annotated Bibliography   183
Index   191

Preface

After thirty years of fossil collecting in the Appalachians, I can count on my fingers and toes the number of times I have seen other collectors in the field. It seems that fossil hunting is not a very popular pastime. Though in most places nice fossil specimens are much easier to come by than good mineral specimens, mineral hunting is so much more widely pursued that the fossil hobby is usually considered a minor specialty within its domain. The truth is, I have long since come to expect that mentioning my fossil collection will send eyes to the ceiling and feet toward the door.

Why is this? I think many people are a little afraid of fossils. Fossils baffle, or challenge beliefs, or seem so strange and alien and ancient that they make one feel a tad insignificant. But fossils are among the most interesting and beautiful objects in nature. Living in ignorance of them is like living in a ruined city without knowing who built it or how they lived. It is true that collecting can be hard, dirty work. But one can set one's own pace, and many localities require no more exertion than a leisurely stroll with open eyes—and a willingness to bend over now and then.

Perhaps a major reason for the general indifference to collecting fossils is suggested by the following scenario: A camper turns up a graceful ribbed shell fossil in the mountains. In my part of the country, there's a pretty good chance it will turn out to be an articulate brachiopod named *Mucrospirifer mucronatus* from the Middle Devonian Mahantango shale. Many people would rather not know what it is than try to digest all of that.

So maybe the problem with fossil hunting is that understanding what one finds involves too many fancy names and fancy theories—too much dependence on the esoteric knowledge that swims in the heads of a few Ph.D.s. But Ph.D.s and other fossil lovers started out as little children with gaping mouths and bulging eyes, fascinated by the mystery of the past and the beauty of its remains.

This is my attempt to write the book I always wanted when I was just such a kid. None of the volumes in libraries or bookstores was able to satisfy my thirst for pictures of fossils or specific collecting information. And since I am still a kid where fossils are concerned, there was nothing to do but to write that book at long last. Since my collecting experience has been mostly confined to the area within easy reach of my home in northern Virginia, the focus of this book is on the collecting opportunities in Virginia, West Virginia, Delaware, Maryland, and southern Pennsylvania. However, most of the fossils described here do occur elsewhere in the United States, and the collecting techniques may be applied anywhere.

In a sense, this effort is a substitute for the career I never had. After studying biology in college, I began graduate work as a paleontologist—a scientist who studies prehistoric life. But I found there was no opportunity on a professional level to cultivate my particular fascination with fossils. I was not interested in specializing in the history of one group of animals, nor the fine points of classification, nor the statistical analysis of ancient populations. I hasten to state that I'm glad so many brilliant men and women have been interested, since I have benefited so much from their work. The professional publications that deal with fossils give one an idea of the richness of the science of paleontology. The opportunities for new discoveries and for greater understanding of the Earth's past through scientific investigation are virtually limitless.

From the scientific point of view, a fossil is a source of valuable information; it can reveal much about the climate and geography of the past. The study of many specimens can reveal details about the history and development of life and the ways and rates at which species of organisms change over time. Fossils can also help locate such resources as oil and coal and certain mineral deposits.

From my point of view, a fossil is first of all a thing of remarkable beauty that still holds something of the past within it. It is a messenger full of mystery that cannot be described with scientific labels; so I draw a picture or write a poem or approach a road cut like a shrine or a cathedral. The awe and exhilaration so many of us feel in the presence of the ocean, or the pyramids, or the Great Wall of China come to me when I find a broken shell that is a million years old. By imagining ancient worlds and identifying with their strange inhabitants, I feel as if I can travel in time. Nothing in nature can seem more mute than the stone impression of an organism whose kind disappeared half a billion years ago, yet almost nothing can be more eloquent if we have the ears to hear. The scientists who listen have heard wondrous things and tell remarkable stories about the past.

But there are many things that fossils say to all of us about beauty and life and death and time. Since there is no better way to "hear" than to find and commune with your own fossils, much of this book describes techniques for collecting and preparing specimens and gives directions to collecting localities I've come to know.

I would like to acknowledge my profound debt to the generations of scientists and amateurs who have made fossils come to life. Their meticulous exploring, collecting, describing, correlating, mapping—and the inspired deductions that they teased from the evidence—have revealed worlds gone by. I am specifically indebted to Dr. Roy S. Sites, of the Virginia Division of Mineral Resources, in Charlottesville. He provided me with useful information and advice, and his interest in my collecting experiences encouraged me to write this guide to fossil collecting.

My editors, Anders Richter, Jackie Eckhart Wehmueller, and Diane Hammond, and the book's designer, Ann Walston, deserve thanks for their patience and many excellent suggestions during the evolution of this volume into what I hope will prove a useful and informative guide to a fascinating hobby.

And finally, I would like to thank my brothers, David and Phil Burns, for lending me specimens from their own collections for the illustrations in this book —and for sharing the adventure of fossil hunting with me.

Part I Understanding Fossils

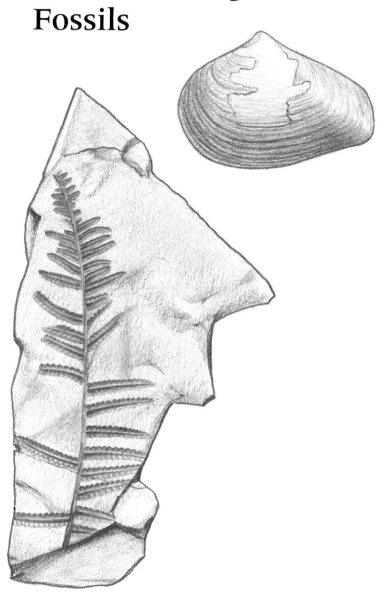

What Is a Fossil?

Exactly what is a fossil? A fossil is any trace of life from the geologic past. It may be a piece of a plant or an animal—leaf, bone, shell, or claw—or an impression of such a piece. It may even be a track or a trail, a burrow or a nest, a footprint. A frequently asked question is, How old does something have to be to be considered a fossil? Scientists generally define fossils as being at least ten thousand years old. This age is not entirely arbitrary but corresponds to the beginning of the current epoch of Earth's history and the end of the previous one, known popularly as the Ice Age. The oldest fossils ever found are 3.5 billion years old, about a billion years younger than Earth itself.

The vast majority of fossils are the remains of marine organisms, because most of the rocks that contain fossils were formed from deposits of sediment that accumulated on the bottom of shallow seas. Less common, but still abundant in some areas, are the fossils of plants and animals that lived in freshwater lakes and rivers or along their shores. The rarest fossils are of organisms that spent most of their time away from bodies of water; their tracks and remains were much less likely to be protected by burial than those of aquatic organisms. Such fossils are usually found where these animals visited bodies of water or where their remains chanced to be submerged and buried after death.

Not all fossils are preserved in underwater sediments. Some are found under volcanic ash or windblown sand, or embedded in ancient tar pits, bogs, ice, or frozen soil. Insect fossils have been discovered in hardened resin, known as amber. Other specimens have been protected from destruction by being in caves or underground crevices.

In order to appreciate the richness of the vast fossil record, full of forms that bear little resemblance to the life of today, we need a basic understanding of Earth's story. Much of this story is told in scientific terms that may seem unpronounceable—may even *be* unpronounceable. But these terms enable us to make sense of the history of life and to begin to understand the fossils we find.

There are two categories of terms to get friendly with: the names of the divisions of Earth's history and the names of the groups of plants and animals whose fossils bring it to life.

Table 1. **The Geologic Time Scale**

	Elapsed Millions of Years	
	Began	Ended
Cenozoic Era	65	
Quaternary period	1.6	
Holocene epoch	(10,000 yrs. ago)	
Pleistocene epoch	1.6	(10,000 yrs. ago)
Tertiary period	65	1.6
Pliocene epoch	5	1.6
Miocene epoch	24	5
Oligocene epoch	37	24
Eocene epoch	58	37
Paleocene epoch	65	58
Mesozoic era	245	65
Cretaceous period	144	65
Jurassic period	208	144
Triassic period	245	208
Paleozoic Era	570	245
Permian period	286	245
Pennsylvanian period	320	286
Mississippian period	360	320
Devonian period	408	360
Silurian period	438	408
Ordovician period	505	438
Cambrian period	570	505
(Precambrian eras)	4,600	570

Note: The Pennsylvanian and Mississippian periods are often combined and called the Carboniferous period.

Setting the Stage: Geologic Time

Nearly all fossils come from the last three eras, or major time divisions: the Paleozoic, the era of "ancient life"; the Mesozoic, the era of "middle life"; and the Cenozoic, the era of "recent life." We are living in the Cenozoic era. Together, these eras represent only about one-eighth of the history of this planet, but they produced most of the fossil record. Fossils found in rocks older than the Paleozoic are rare and are usually poorly preserved.

Each era is subdivided into smaller units of time, known as periods. Memorizing the periods is a help in getting a feel for the stages and progressions in the story of life. In fact, each period is like a chapter from a great epic, with its own characters and plot. In a sense, Earth was a different planet in each of them, and you may find such concepts as the Devonian world or the Cretaceous seas useful tools for your imagination. Table 1 shows the approximate numbers of years since the beginnings and endings of these eras and periods, as well as the subdivisions of the two Cenozoic periods, which are known as epochs.

Almost invariably, when I show a fossil to noncollectors and say, "It's a hundred million years old," they respond with, "How do they know how old it is?" I then realize there is no way to answer without sounding very technical.

The fact is, these dates are based on radiometric dating, the measurement of "before" and "after" proportions of unstable elements in rocks. These unstable elements undergo transformation because of radioactive decay at fixed and known rates. The age of certain rocks can be determined by measuring the proportion of the original unstable elements and the more stable products of their decay. This principle of constant rates of radioactive decay is also used in atomic clocks and is very precise and reliable.

Having explained that, I usually see a blank stare or glazed eye that seems to say, "Oh, I see, it's just a theory," which, of course, it is not. But if this line of evidence isn't convincing, another may be.

For example, we might consider how long it takes to accumulate thousands of feet of layered sedimentary rocks, each layer preserving different fossils in an orderly succession, which is globally consistent. In many places, one can distinguish and count the progressive coral reef communities or the floors of mature forests, one on top of the other in the rock layers. Each of these environments existed for many years before it was buried. Estimates of the age of fossils based on sedimentation rates ranged into the tens and hundreds of millions of years, long before radiometric dating was invented. But see for yourself—you can count the years in the rocks themselves.

Knowing the Actors: The Many Names of Fossils

The second list of terms to learn is the names of the different groups of organisms represented in the fossil record.

Plants and animals are classified into a hierarchy of groups, starting with broad categories called *kingdoms*. The plant and animal kingdoms are divided into smaller and smaller groups, each level reflecting a closer relationship between its members. Kingdoms are divided into *phyla*, which are divided into *classes*, which are divided into *orders*, which are divided into *families*, which are divided into *genera*, each of which contains at least one *species*. (Scientists also recognize many intermediate groups, such as *superfamilies* and *subphyla,* but we need not be concerned with those here.)

For an example, let's classify that brachiopod our camper discovered. We know it belongs to the animal kingdom and to the phylum (singular of phyla) Brachiopoda (all phylum names end with *a*). Further classification requires a close inspection of the internal structure of the shell—which may not be well preserved or easy to see—so we'll rely on matching its external features with a picture in a guidebook. We find that it belongs to the class Articulata, the order Spiriferida, and the family Mucrospiriferidae (the *-idae* ending is standard for families). The genus (singular of genera) name is *Mucrospirifer,* and the species name is *mucronatus* or, more properly, *Mucrospirifer mucronatus.*

Learning these categories and names may seem less daunting if we remember that they are just names, like *Smith* or *cat* or *koala bear.* They don't contain any great ideas or secrets known only to scientists. They mean something, of course, and it's nice to find out what, but how much do you gain by knowing that *brachiopod* means arm-foot? Especially when what looks like an arm isn't really a foot.

The sentence "The man walked the dog" could be translated "The *Homo sapiens* walked the *Canis familiaris.*" We don't use those fancy scientific names because we have simpler, more familiar ones. But familiar names aren't available for most prehistoric beings, and even when they are, problems arise. For example, most people recognize several types of bivalved mollusk: clams, oysters, scallops. A few people know the common names of more living types; but what do you call an extinct form? On page 85 is an impressive specimen called *Praecardium multiradiatum.* Formidable name, but what else can you call it? How about a pelecypod? That's the inclusive name for nearly all bivalved mollusks, living and fossil, and it gives you something to call a *Praecardium* before you know it's a *Praecardium.*

Similarly, snails are called gastropods and sand dollars, echinoids. We could muddle through here without the Greek and Latin, but what about extinct or obscure forms like trilobites, ostracodes, crinoids, and nautiloids? And remember, these terms are recognized throughout the world, regardless of their common names in the local language.

Dinosaurs prove that the names needn't be a problem; how many five-year-olds do you know who can tell a tyrannosaurus from a triceratops? Probably more than there are categories of fossil groups to learn. Before too long, you'll find yourself savoring names like *Praecardium multiradiatum,* as well.

Part 4 of this book introduces nearly all of the major groups of plant and animal fossils you are likely to find, with brief descriptions and references to representative fossils that are illustrated in Part 3. Terrestrial vertebrate fossils are not discussed, as they are not only rare but present a whole new menu of jaw-breaking names. Robert L. Carroll's *Vertebrate Paleontology and Evolution* (W. H. Freeman, 1988) lists nearly all groups of fossil vertebrates for those who wish to pursue this kind of collecting.

Underwater life of the Silurian period. The scene is dominated by three straight-shelled nautiloids. Similar species from the preceding Ordovician period had shells as long as twelve feet. The soft parts of these animals are reconstructed to resemble those of the living chambered nautilus, the only surviving nautiloid, as these anatomical details are not preserved in fossils. A trilobite (left) rests on the ocean floor near two crinoids, and a horn, or rugose coral, with its polyp fully extended, is on the right.

One of the benefits of mastering the different groupings is that you become more aware of the diversity of *living* organisms, especially when you visit the beach. A shell will never be merely a shell again.

How They Got There: A Brief Geologic History of the Area

During the last century and a half, some great minds have contemplated the question of how the rocks, fossils, mountains, valleys, and other geologic features have come to be distributed as they are. But no truly coherent explanation was possible before the acceptance of plate tectonics—the realization that Earth is a dynamic planet whose surface is divided into distinct crustal, or tectonic, plates, which have changed position through geologic history and which are moving still. These movements are caused by convection currents in the Earth's mantle, deep beneath the crust, which bring new material to the surface in the form of molten rock, causing the crustal plates to diverge from each other in some places and to converge in others.

The rates of movement are very slow relative to our sense of time; one inch per year would be a very good clip. But when these rates of change persist over many millions of years, the configuration of the Earth's surface is completely transformed.

As the margins of plates move against each other, oceans may disappear as the denser oceanic crust is driven under the less dense continental crust. The impact may create coastal or off-shore mountain ranges, and the heat caused by friction as the crustal margins move against each other may melt rocks and create volcanoes. The island of Japan and the volcanoes that ring the modern Pacific Ocean were formed by collisions of this kind.

Vast mountain ranges may also be thrust up by the collision of continental crust material on converging plates. As they are of approximately the same density, neither of the colliding continents will dive beneath the other; they will crumple together as mountains. The Himalayas are still growing higher as the piece of continental crust known as India is forced against the Asian continental crust.

Forces and events such as those described above have played their parts in the geologic history of the area covered by this book and have left their marks. Many of the rocks in the Appalachian Mountains, for example, have been severely folded and their fossils destroyed or distorted by the pressures brought to bear in the collisions of crustal plates.

A detailed accounting of the history of even a small area is far beyond the scope of this book, but a thumbnail sketch of the story seems worthwhile. Since virtually all of the fossils to be found in the region come from rocks that are Cambrian or younger in age, we'll skip the first four billion years of the story.

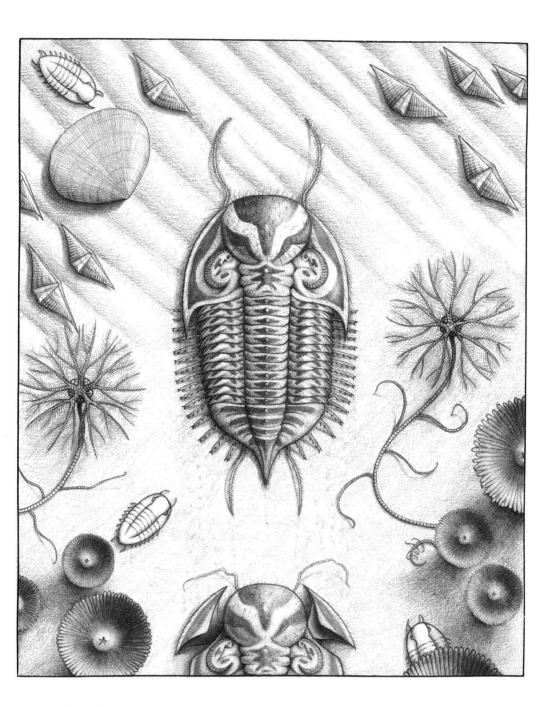

Marine life of the Devonian period. In the center is a large trilobite, five inches long. Like all arthropods, trilobites molted periodically in order to grow, as their shells were incapable of much stretching. This individual has just left his old exoskeleton (bottom). Surrounding him are brachiopods (top right and left), horn corals seen from above (lower right and left), crinoids on either side, and three much smaller trilobites of another species. Note the antennae and legs of the trilobites; these are very rarely preserved in fossils.

The Paleozoic Era

Five hundred and seventy million years ago, in the beginning of the Cambrian period, what is now eastern North America was separated from northern Europe by an ocean, as it is today. But this barrier was not the Atlantic Ocean. In the Cambrian and early Ordovician periods, much of the area was covered by the warm tropical waters of the Iapetus Ocean, which flooded the low-lying margin of North America. The region was then situated near the equator (and it would remain in tropical latitudes throughout the Paleozoic). The early Paleozoic rocks of the Mid-Atlantic region reflect the changes brought about by the gradual closing up and disappearance of the Iapetus.

Continental seas were widespread in North America during most of the Paleozoic era. They are quite different from oceans proper and represent the flooding of continental crust, like the Chesapeake Bay and the water above the continental shelf. In contrast, most ocean waters are situated over oceanic crust and are much deeper. Today's oceans average about two miles in depth as compared to typical continental sea depths of a few hundred feet or less.

The nearby land areas in the Cambrian period were low lying and contributed relatively little sediment through erosion, so the deposits that accumulated offshore were composed largely of the remains of the wide variety of marine animals that thrived there. The shells of most marine organisms are made of calcium carbonate, and the rocks they form are called limestones. Limestones from this time are widespread in the Mid-Atlantic states today, especially in the Shenandoah Valley and southeastern Pennsylvania.

By the middle to late Ordovician period, local geography changed dramatically as the European crustal plate moved toward North America. As it approached, the leading edge of the dense oceanic crust of the Iapetus Ocean floor dove under the continental crust of the North American plate, and the Iapetus gradually shrank. These movements and the collision of plates that they caused created volcanoes and a mountain chain along the eastern margin (relative to the modern position of the continent) of North America. A mountain-building event such as this is termed an *orogeny;* this episode is known as the *Taconic orogeny.* The erosion of this new land produced quantities of mud and sand, which accumulated along its shores. Marine life continued to thrive in the shallow waters around the mountains, and many of the sediments became shales and sandstones, which yield abundant fossils in the region today.

By the late Silurian period, Europe (and Greenland) had reached North America, and the Iapetus Ocean was no more. The mountains were largely eroded away, and widespread limestone formation was again occurring in the area. But the middle to late Devonian period saw a renewal of mountain building (the Acadian orogeny), as the European plate was pressing against what is now northeastern North America. Once again, mountains were thrust up and quantities of mud and sand were deposited in the surrounding continental seas, in-

Labyrinthodont amphibians of the Pennsylvanian period. Shown here in a hypothetical mating ritual, two five-foot-long males squabble over a female. A seed fern towers above them. The term *labyrinthodont* refers to the complicated structure of their teeth, in which the enamel is so complexly folded as to resemble a labyrinth, or maze, when viewed in cross section. Fossils of these animals may be found at locality 38.

cluding those that covered most of the Middle Atlantic states. The northern Appalachian Mountains were formed at this time.

In the late Paleozoic era, luxuriant forests grew up along the shores of rivers and the continental seas. As sea level rose and fell, the forests were often buried by marine sediments. These remains have become the coal deposits of West Virginia, Pennsylvania, and southwestern Virginia. By the end of the Paleozoic era, the northern European crust was pressed against northeastern North America, and what is now the African crust had collided with what is now the southeastern United States. This continental collision closed the ocean that had separated Africa from North America and built the southern Appalachians (the Alleghany orogeny). The mountains that existed in eastern North America at this time may have rivaled the modern Himalayas in height. Their remnants—separated by subsequent seafloor spreading and reduced in size by millions of years of erosion—stretch from Arkansas to Norway.

As the mountains continued to rise, the land areas around them were also uplifted and the continental seas in the interior of North America receded. Before the Mesozoic era began, eastern North America was landlocked. Indeed, almost all of the continental crust in the world had come together to form one supercontinent, which we call Pangaea. This event caused a dramatic decline in shallow-water marine environments, which resulted in the extinction of many forms of marine life.

The Mesozoic and Cenozoic Eras

Toward the end of the first period of the Mesozoic era, the Triassic, the dinosaurs made their appearance, and the supercontinent Pangaea was beginning to split up. Europe and Africa, which had spent so long—and had caused so much commotion—coming to rest against North America, were now starting to move away. A few false starts occurred to the west of the present coastline, producing a series of rifts, or cracks, in the Earth's crust. Today, a narrow band that parallels the Appalachians contains intermittent outcrops of red shales and sandstones deposited around and in the freshwater lakes that filled these rifts. Igneous rocks, formed from the molten rock that oozed from the wounds of the Earth's crust as it tried to split apart, also occur in this belt. (A similar situation exists in the rift valley of modern eastern Africa, where the widening Red Sea marks the opening of an ocean of the future.) Eventually, one rift "took" and continued to widen, giving birth to the Atlantic Ocean—which is still growing in size today through "seafloor" spreading.

After all the dramatic upheavals of the Paleozoic and early Mesozoic eras, eastern North America has been a model of tectonic stability, with no mountain building or plate collisions. The eastern margin of the continental crust has ac-

Marine life of the Jurassic period. A plesiosaur (his long neck sweeping out of view at the top and back into view at the left) pursues a school of fish. Plesiosaurs, which were reptiles but not dinosaurs, swam in the Mesozoic seas of the region. Their fossils are occasionally found locally. The specimen here is about thirty feet long. His prey are early members of the group of fish known as the ray fin fishes, which includes the majority of modern species.

cumulated sediments in much the same way as it did in the beginning of this story, in the Cambrian and early Ordovician periods. One major difference is that North America gradually wandered into cooler latitudes and, except for its extreme southeastern margin, is no longer a good setting for limestone deposition. But plenty of mud and sand have accumulated from the Cretaceous period on, much of it yet to be consolidated into rocks, and these sediments contain a rich variety of fossils.

Less than two million years ago, at the beginning of the Quaternary period, the polar ice caps began to grow. Ever since that time, the ice has advanced and retreated by turns, reaching the northernmost fringes of our area at its greatest extent. When the ice caps are extensive, they contain enough water in the form of ice to lower sea level, so the position of the shoreline is partly dependent on their size. Fortunately for the fossil collector, the caps are larger and sea level is lower today than it has been during most of the time since the Atlantic Ocean began to open up in the Mesozoic era. This means that many fossiliferous marine sediments are now on dry land. Of course, even more sediments were exposed when sea level was at its lowest during the time of maximum ice cover.

The preceding account is a great simplification and barely hints at the enormous complexity of the geologic history of the region. But the main point has been made—things are not as they once were.

Getting to Know the Sedimentary Rocks

The erosion of mountains formed by tectonic forces, and of land surfaces in general, produces mud and sand and gravel, which accumulate in oceans, inland seas, lakes, riverbeds, and on continental shelves. These sediments may eventually become shales, sandstones, and conglomerates (gravel stones). Fine-grained sediments, like mud and silt, are turned into rocks by being compressed under the weight of overlying deposits and water. Coarse-grained sediments, like sand and gravel, are hardened when the grains are bound together by mineral cements such as dissolved silica (quartz is a common form of silica) or calcium carbonate, which precipitate out of solution and set up after having been spread throughout the deposits by groundwater.

If living things are present when these sediments gather, and if conditions are right, remains or traces of such life may be preserved as fossils in the rocks. Fossils are also frequently preserved in limestones, which are formed from the accumulation of organic debris, such as the calcium carbonate shells of organisms, or from the precipitation of excess calcium carbonate in seawater. If enough mud or sand is present in the sediments, shaley or sandy limestones are formed. If there is more mud or sand than calcium carbonate, the rocks are known as calcareous (or limy) shales or sandstones.

Terrestrial life of the late Jurassic period. A herd of giant plant-eating sauropod dinosaurs watches in the distance as three twenty-five-foot-long meat-eating theropods close in on their prey (not shown). The fact that some dinosaurs engaged in herding behavior has been well established, but it is not known if large theropods such as these hunted together or shared their meals. Fossils of both of these types of dinosaurs have been found in the Mid-Atlantic region but are not common.

Understanding Fossils 15

Sandstones can be divided into two basic types, either of which may be fossiliferous (fossil bearing). These are known as *mature* and *immature* sandstones. Immature sandstones are made of sand deposited during periods of rapid erosion of the land, as when mountains were building nearby. Besides the very resistant quartz grains, many other less resistant minerals were deposited and buried before they could be broken down by weathering. These minerals give these sandstones their characteristic green, gray, red, brown, and purple colors.

Mature sandstones, on the other hand, consist almost entirely of quartz grains and are usually white or light gray. They are formed from deposits that accumulated gradually near low-lying, stable land areas. Typically, these deposits are moved about and are reworked repeatedly before they are buried and turned into stone—long after the less resistant minerals have been broken down and weathered away. The sands found on the modern beaches of the Mid-Atlantic coast are very low in minerals other than quartz and may well become mature sandstones some day.

The shells of some microscopic marine organisms, such as diatoms and radiolarians, are made of silica. The rocks that are formed from deposits of these shells are known as cherts. Cherts are commonly found as layers or nodules associated with limestones and often preserve fossils.

Since different kinds of sediments occur in different environments, the rocks that contain fossils can tell us much about the conditions in which the organisms lived. For example, shallow tropical seas are usually necessary to support enough marine organisms to form organic limestones. So when you find a fossiliferous limestone outcrop, you know the rocks were probably formed from sediments laid down in a warm, shallow sea.

Because shales and sandstones are made from the products of the erosion of land surfaces, land must have been near to the places where most of these sediments accumulated. Different kinds of sand and mud are produced by erosion of different kinds of land, making it possible to say whether a particular shale or sandstone was formed near a volcano, a glacier, or gently rolling hills. The amount of information that a scientist can get about the origin of rocks from the careful study of a rock exposure is staggering. Once you recognize what rocks say about the past, you begin to get a clearer idea of the geologic history of your area.

Let's go back a bit to the right conditions for making fossils. Many factors work against the fossilization process, so many that the odds against any particular bone or shell becoming a fossil in the future are extremely high. First of all, a host of organisms make use of the remains of dead organisms. Even apparently useless leftovers like clam shells are soon riddled with holes made by sponges, which bore into them. Marine shells are often simply dissolved by the sea or reduced into sand by waves and currents. Even if the would-be fossil is lucky enough to get buried by sediments that protect it from destructive agents, there is no guarantee that it won't eventually be dissolved by groundwater or that the protecting sediments won't be washed away or melted by geologic forces.

Flightless birds of the Eocene epoch. These giant birds, members of the family Diatrymidae, stood over six feet tall. Such birds were widespread in the early Cenozoic era, after the extinction of the dinosaurs and before large land mammals had evolved. Their fossils are found in the western United States and in New Jersey, so it is likely that they lived in the Mid-Atlantic region, as well.

One of the consequences of the high odds against preservation is that only a tiny fraction of the species of animals and plants that have lived on Earth are represented in the fossil record. Few species survive more than five million years. There are nearly two million species known to be alive today, but fewer than 150,000 species of fossil plants and animals are known—about one-tenth of 1 percent of all the species that have lived. New fossil species are still being discovered, but clearly, most of the kinds of organisms that inhabited the prehistoric worlds became extinct without leaving a trace.

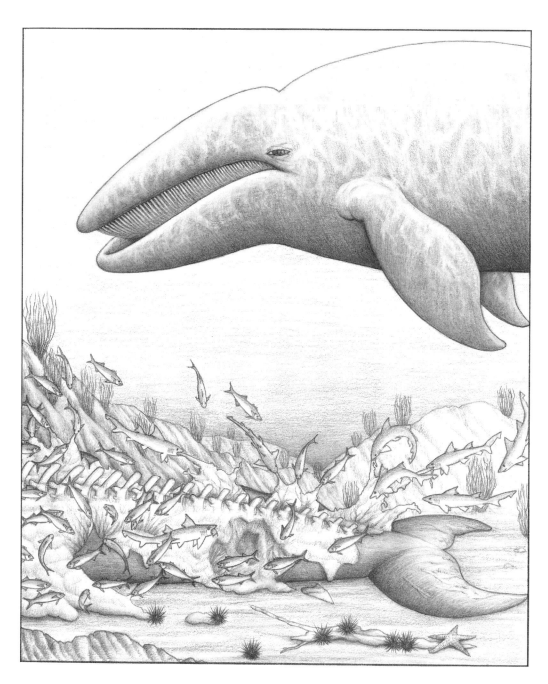

Aftermath of a shark attack, Miocene epoch. Modern sharks do not prey on full-grown whales of the larger species, but prehistoric sharks of the genus *Carcharodon* were so enormous, they could have done so. The occurrence of whale and shark fossils in the same sediment layers, plus scars in whale bones apparently made by sharks' teeth, indicate that these sharks did prey on whales. The whale shown swimming is a *Cetotherium,* one of the baleen whales. The carcass of the whale that was attacked has attracted various scavengers, including spiny dogfish (small sharks). Both *Carcharodon* and *Cetotherium* fossils are found in the Miocene deposits of Maryland and Virginia.

Part II How to Collect Fossils

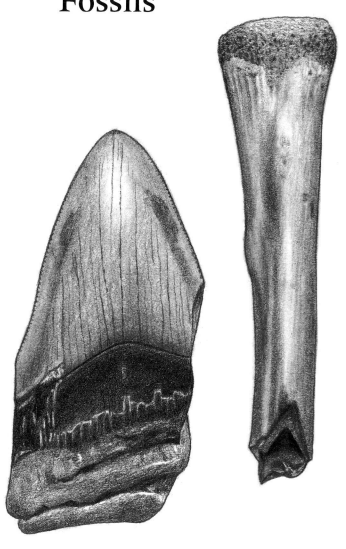

How to Start

What you need to collect fossils is, first, a locality—a beach, a road cut, a quarry, a ditch, or a pile of rocks—where fossils can be found. If you are a reasonable sort of person, you'll visit the places described in collector's guidebooks or consult with a rock and mineral club. If you are an egomaniac like me, you'll wander from rock pile to road cut as if you've just discovered a strange new planet, and see what turns up. (But if you choose my way, make sure you're in country where the rocks are sedimentary in origin—shale, sandstone, or limestone—so you have a chance for success.)

A sort of compromise approach to finding fossil localities is to predict their occurrence from information available in geologic publications, particularly field reports and geologic maps. These may be obtained from state geological surveys (or equivalent organizations) and the U.S. Geological Survey. (Addresses are listed in the appendix.) These publications will provide you with all the information you need to find fresh localities—and to fill every corner of your house with fossils.

The key to successful collecting is recognizing the richest rock formations. A formation is a body of rock, usually widely distributed, that is distinguishable from the rock layers above and below it by its composition and the fossils it contains. Some formations are utterly devoid of fossils, while others may be virtually made of them. Formations are given names, usually taken from localities where they are particularly well exposed or where they were first studied (e.g., the Martinsburg formation, the Calvert formation). Sometimes formations are referred to by names that reflect the dominant rock type, such as the Needmore shale, the Oriskany sandstone, the Lincolnshire limestone.

Related formations are often classified in groups (e.g., the Helderberg group comprises several early Devonian formations). Distinctive parts of individual formations are called members or zones (e.g., the Wymps Gap limestone member of the Mauch Chunk formation and the *Orthorhynchula* zone of the Martinsburg formation, the latter being named for a genus of brachiopod).

A geologic map is an especially useful tool for finding where fossil-rich formations are exposed. What the map actually shows is the geographic area where each formation comes to the surface (or is overlain only by soil, vegetation, or humanmade structures). The map's legend will include a list of the formations in the mapped area with brief descriptions of each, including whether or not it contains fossils. Generally speaking, formations of Pre-cambrian rocks, igneous rocks (made from molten material), and metamorphic rocks (drastically altered by heat or pressure) may be considered to be without fossils.

An excellent way to find a fossil locality is to look for rock exposures (road cuts, quarries, natural outcrops, etc.) in an area where a fossiliferous formation comes to the surface. Plan a driving or hiking route through promising areas, and check every accessible outcrop. In places where an especially productive formation is exposed, nearly every boulder, stone, and pebble can contain fossils, so a geologic map holds the key to finding an almost limitless number of fossil localities.

Geologic maps may be purchased separately; more often they come with field reports and bulletins published by the national and state surveys. The bulletins are extremely useful in themselves, since they provide a list of the rock formations in the study area, a discussion of their geologic significance and approximate age, and a description of the kinds of rocks and fossils found in each formation. They also identify localities with exceptional exposures.

Once you've found a source of fossils, you're in business. Look at it as you would an unexplored beach, or forest, or underwater reef. The best way to learn to read the rocks and know the fossils is to see your locality as a window to a living world and then get to know its inhabitants, or at least the ones who left traces behind.

Identifying What You Find

Figuring out what kinds of fossils you've found can be a fascinating process—like a good mystery or riddle. It can also be frustrating. You'll find your powers of observation challenged and strengthened. The first step in identification is to recognize what major group of organisms a specimen represents. Some types, like complete trilobites, are easy to distinguish, but it will take longer to learn to tell a gastropod from a coiled cephalopod, or a bivalved brachiopod from a bivalved pelecypod.

Though many types of fossils occur, the vast majority of specimens can be readily assigned to a handful of major phyla and classes, which you will learn to recognize very quickly. The names of many genera will soon become familiar, as the guidebooks you use to identify your specimens give the generic names. In most cases, individual species can be identified only by experts. Exceptions include distinctive, widespread species, such as the trilobite *Phacops rana*. But unless your guidebook identifies its illustrated fossils to species level and also states that the species occurs in the formation your specimen came from, it's best to be content with knowing only the genus. Orders and families will take a bit longer to recognize, since identification to genus level is possible (using the guidebooks) without knowing which of these categories your fossil belongs to. However, a familiarity with them will help you to understand the interrelationships and evolutionary histories of your fossils.

Once you're certain which formation your specimens come from, you can match most of your finds with the list of fossils in the field reports, using one or more of the guidebooks to fossil identification. But beware! Geologic maps are not always easy to read, especially in areas with complex geologic structure or when the maps cover a large area in a small scale. I misidentified the formation for the first really good locality I ever found and did a respectable job of misidentifying every fossil I brought home before I discovered my error. Almost always, each formation will contain one or more guide or index fossils, types that are common in the formation and that occur only in one or two formations. These guide fossils are mentioned in the field reports.

Success in fossil identification depends largely on the number and quality of your guidebooks. Give the Annotated Bibliography a close look; a half dozen carefully chosen volumes should be all you'll ever need to identify nearly every fossil you find. The fact that, even after years of collecting, you'll continue to find a few fossils that defy easy identification only adds interest to the hobby.

Sizing Up a Road Cut

Just as the key to finding fossil-rich rocks is to identify the most productive formations, the key to the best hunting at a locality is to find the best layers within the exposure. First, give the loose rocks and debris at the base of the outcrop (known as scree) a thorough going over. These rocks are a sampling of the layers above them. At some places, this will be the only rock that is easily or safely available. I spent years concentrating on such rockfall due to sheer laziness and still accumulated a vast collection. If you want to find the real prizes regularly, you will want to mine the richest layers where they sit. But this can be backbreaking, dangerous work, and it may take an advanced case of fossil fever to get you to attack the cliff face itself.

For years I tried to discover how to recognize fossiliferous road cuts from a moving car. I have learned to identify a few productive formations, but my attempts to correlate color of rock, thickness of beds (or layers), or texture of rock with fossil content have not been productive. A few generalizations and observations about different types of sedimentary rocks are worth making, however.

First of all, red beds are usually poor in fossils. Red beds are the red or purple shales and sandstones usually formed in nonmarine environments, such as in river deltas or floodplains. Sometimes they contain a few fish scales or teeth or the extremely rare complete animal; trace fossils, the markings left behind by plants or animals, may occur. Triassic red beds that accumulated near rift valley lakes contain fossils, especially dinosaur footprints, at a few localities. But these are exceptions, and as a general rule, hunting for fossils in red shales and sandstones will produce little or nothing. Other clues to past worlds, like ripple marks and mud cracks, are fairly common, however, and worth noticing.

Though limestones are often full of fossils, they may be very difficult to collect from. Often they are tough and will not break in ways that expose recognizable fossils. However, the larger fossils are sometimes more resistant to weathering than the surrounding rock (or matrix) and may stand out in relief on the surface. Some of the finest specimens of this type are obtained from limestone in which the calcium carbonate of the shells was, by a poorly understood process, gradually converted in place into silica (quartz). This metamorphosis is a boon to fossil collectors, because limestone dissolves much more readily than silica when exposed to rain, which is naturally somewhat acidic. The result is that the silicified fossils are gradually released from the surrounding rock as it is dissolved away and may be collected with ease. Fossils of this type may also be found in soils derived from the weathering of such limestones.

Most limestones contain considerable amounts of sand and mud mixed with the calcium carbonate, which is their main constituent. As fossil shells are usually composed of relatively pure calcium carbonate, they are sometimes poorly integrated into the matrix. This may cause the rocks to break in ways that expose the fossils within them. Sometimes the sand or mud may form discrete layers and contain fossils that are more visible than similar fossils in layers that are more calcareous (higher in calcium carbonate). Chert, which is commonly found with limestones as nodules or in layers, often bears impressions of the fossils contained in the surrounding limestone.

Limestones come in all colors, but most of the formations in the Mid-Atlantic states are shades of gray. In the Shenandoah Valley, the rounded, light gray boulders that crop out in fields and pastures are limestone. Many of the large quarries throughout the eastern Appalachians are in gray limestones, some of which are extremely fossiliferous.

Most of the Paleozoic shales and sandstones in the eastern Appalachians are much tougher than the fossils they contain. These fossils are usually made up of a crystalline form of calcium carbonate known as calcite, which dissolves readily on exposure to the elements. For this reason, many of the best fossils in shales and sandstones are ghosts—cavities in the rocks left behind by shells that have been dissolved away by rain and groundwater. Many times, I have split open a chunk of fresh shale to find the interior sparkling with the calcite cross sections of irretrievable fossils.

The fossil content of such rocks is best appreciated as the impressions and molds left behind after the shells have been weathered away. These consist of external molds, which may preserve fine surface details and, in the case of hollow fossils, internal molds, also known as casts, or steinkerns. Sometimes the ghosts have been filled by new minerals precipitated by groundwater. These materials may or may not be more resistant than calcite or the remaining rock.

Besides weathering rapidly, calcite is also very brittle, much more so than most shales and sandstones, so collecting in these rocks depends largely on the results of the weathering process and the way the rocks happen to break. Bedding planes, which represent pauses in deposition during which shells could accumulate, separate one rock layer from another. They are easily exposed, since they represent zones of weakness in the rocks and are frequently covered with fossils.

Many shales have a tendency to split into thin sheets, sometimes paper thin. Such rocks are called *fissile*. This property can make the rocks easy to break apart while you are searching for fossils. But sometimes the fossils themselves also split into sheets. When collecting from such rocks, have some glue handy.

Once you've found a fossiliferous rock you enjoy collecting in, find out where it occurs in your area. In many formations, distinctive layers, or beds, persist throughout the formation over a wide area. An excellent example is the highly fossiliferous siltstone found at the top (the youngest layers) of the Devonian Mahantango formation. When you identify such beds, you can quickly home in on the best collecting at a locality. Sometimes, only one thin layer will be productive, and the sooner you realize that, the less time you'll waste on barren rocks.

Combing the Ancient Beach

In contrast to mountain rock exposures are the waters of coastal plains—streams, rivers, bays, and the ocean—where the tides and floods do the work for you, washing away the muck and leaving the shells and sharks' teeth shining at your feet. Maybe for this reason, many more people go beachcombing for fossils than would ever lift a rock hammer.

The same principles of locating the most productive layers of sediment apply at the beach as at a road cut. Look for auspicious associations. For example, the teeth of the giant shark *Carcharodon* tend to occur along with whale bones: when you find a lot of bones, watch for the teeth.

Before you go to the beach, be sure to check a tide chart. Most fossils are found in the intertidal area, so high-tide pickings can be slim. Weather conditions are also important. High winds can raise big waves and cloud the water. A wind from seaward might even prevent the tide from going out. Hip waders or chest waders, like those used by surf fishermen, are a great help for exploring in shallow water. Even if your collecting is done ashore, you will probably need to cross some streams. In tidal areas, the tide might catch you; the passage of time can be tricky when you're engrossed in fossil hunting.

The best collecting is on a winter beach after a storm, at low tide, on a calm day, in bright sunshine. Violent winter storms sometimes throw big teeth and bones and chunks of fossil-rich sediment high up on the beach, stealing the sand as they go and thus making the fossils easier to reach and to see. The smaller, lighter fossils are more easily reclaimed by milder waves.

Beach collecting for sharks' teeth and other vertebrate specimens requires creative looking. Often, imposters like broken glass, leaves, twigs, and chips of stone will distract you. Sharks' teeth are frequently darker than other fossils and most pebbles, a phenomenon emphasized by shallow water. Thus, hunting in shallow water is often fruitful. On the other hand, some subtleties of sharks' teeth are obscured by water: the distinctive texture of the root or the polish of the crown. Also, the optic effects of searching through the looking glass of water can make you dizzy. It may be best to switch between stalking the water and stalking the land.

By the way, stooping will increase your take enormously: The closer your eyes are to the fossils, the more you will see. A walking stick is recommended, for support as well as for turning over potential fossils, especially those in icy water. Wearing rubber gloves or carrying a towel to dry off your hands after every dip are good ideas for winter collecting.

What You Will Need in the Field

One reason fossil hunting may be less than the rage is that it gets no commercial promotion since there is so little equipment to buy. The following are the only essentials.

A good rock hammer is a must for collecting fossils from rocks. Get one with some weight, but not so much that you can't wield it gracefully. It may not seem like a very subtle instrument, but you'll soon learn to use it like a precision tool to coax the rocks apart. It doesn't matter much whether it has a pick (pointed) or chisel (flat) end. The pick end may be better for hard sandstones and limestones and soft matrix, while the chisel end works better for most shales. But either kind works well enough in nearly all situations. The only real exception I've found is very hard, thin-layered shales, which are much easier to split cleanly with a chisel end.

Have an assortment of cold chisels or masonry chisels handy. A good selection would consist of three chisels: one chisel four inches long by a quarter inch wide, another six inches long by a half inch wide, and the third twelve inches long by one inch wide. *Do not use wood chisels;* they may produce flying splinters of metal when used on rocks.

If you're an eager beaver like I've become, a crowbar about two and a half feet long is excellent for prying out rock layers. If you're even more of an eager beaver, try a pickax or sledgehammer, but remember, the more destructive force you apply, the greater the risk of destroying fossils—or yourself.

Make sure to use heavy gloves when you're working in all but the softest of rocks, and beware of sharp shells in soft matrix. I never feel I've been fossil hunting until my hands have at least two cuts or blisters; they can't be avoided—but gloves will help keep the damage down. Another must is goggles or glasses. I've

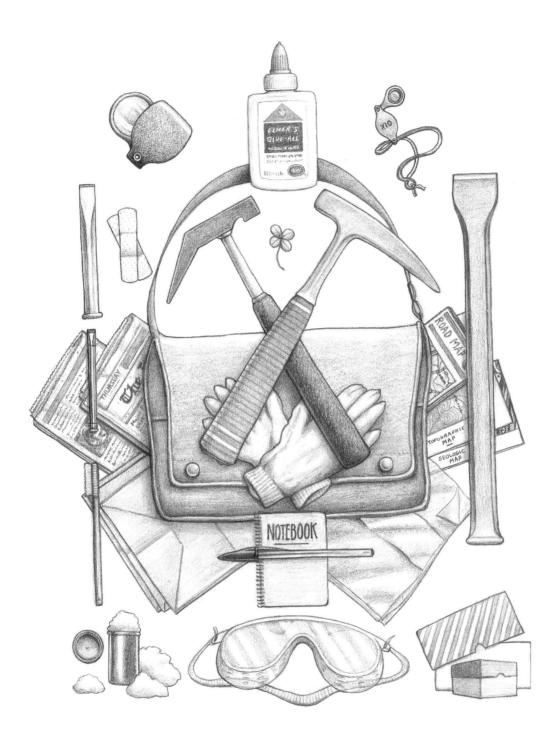

Field gear for the well-equipped fossil collector. Hand lenses (top), Elmer's Glue-all, various sizes of masonry chisels (or cold chisels), bandaids (and other first-aid supplies), a chisel-end and pick-end rock hammer, a rock bag, an assortment of maps, newspapers for wrapping specimens, paper bags, plastic bags, an old toothbrush, a notebook and pen, some cotton, small containers, gloves, safety goggles, and a little bit of luck.

had many a shard of stone glance off my spectacles; which could have done serious damage to an unprotected eye.

Elmer's Glue-all is one of the fossil hunter's truest friends; probably half my collection is held together by it. Some fossil pieces are best repaired in the field for safer transport, so have your glue handy.

A hand lens can be a useful tool. Besides revealing details on the surface of specimens, it can also reveal specimens that are too small to catch the eye. Paleozoic rocks often contain fossils of tiny marine organisms called ostracodes, most of which are only one or two millimeters long and thus are easily missed without a magnifier. I suggest carrying two hand lenses into the field: (1) a small, folding ten-power lens and (2) a two-to-three-inch diameter lens that magnifies objects three or four times. Some collectors are so enthralled with what magnifying glasses allow them to see that they specialize in collecting minute specimens.

Many collectors use sieves for shark's tooth collecting. Sieves are easily constructed from a square wooden frame and a wire screen of a mesh size appropriate to the minimum size of the specimens you are seeking and the maximum grain size of the gravel or sand being processed. Use of a sieve will reveal a quantity of fossils that would have escaped notice otherwise.

A rock bag is a good way to carry your equipment into the field and your specimens out again. Backpacks and tote bags may be used. Once I go to work, I usually lay my rock bag aside as a sort of base camp, where I put fossils as I find them so that I don't have to carry them around any more than necessary. Make sure your base camp is in a place you can find again easily and where it is safe from falling rocks and rising tides. I've spent a lot of time trying to refind my fossils because I broke one or both of these rules. Plastic bags are excellent for beach collecting and for certain hard rock localities where the fossils come out whole or are especially durable. Carry a few; you will probably find one or two prizes that you won't want rubbing shoulders with the others.

Newspaper is excellent for wrapping fossils individually before putting them carefully into your rock bag or, back at the car, into paper bags or cardboard boxes. I take several small boxes, nested in each other to save space, and fill them up as I go. Small boxes or bags are preferable, because they limit the weight on the bottom specimens and keep fossils from different localities separate. Some smaller containers and bits of cotton may also come in handy for tiny and very delicate finds.

Sometimes fossils are best collected with a camera, especially where delicate specimens stand out in relief on large boulders or on the surface of rock layers. You may need a close-up lens to allow you to focus on small fossils. A plus-three detachable lens (at a cost of around ten dollars) for my thirty-five-millimeter camera has worked well for me. Pictures of the collecting localities or of particularly rich parts of them may be worth taking. Besides being reminders of successful outings, they can help you orient yourself when you return.

Make sure to have a good supply of road maps, and keep a notebook handy so you can record the location of the rock exposure and where you found the best material. You may think you can find that road cut again, but your memory may let you down. As many as twenty years have elapsed between visits to some of my localities, and I never would have found them without my notebook.

And (sorry L.L. Bean) that's it! Except for maybe an old toothbrush to clean up your finds as you find them.

Respect Your Road Cut

A few points about how to treat your locality and how your locality might treat you. Whenever possible, *ask for permission to collect*. Land ownership may be difficult to determine at road cuts, but permission should be obtained whenever possible when the road cut is on private property. Never cross fences without permission and respect No Trespassing signs. If you do receive permission to collect, damage the rocks as little as possible so that the owner will be well disposed to future requests from you or other collectors. Remember that there is virtually nowhere that you are entitled to collect, but you can collect almost anywhere if you show some respect and common sense.

Be alert for teetering heaps of rock. Though an outcrop might seem like a permanent structure, those Falling Rock signs you see along the road are telling the truth. I have been shocked more than once to realize that the enormous block of stone I was about to stand on or start hammering on was just waiting for a slight breeze to send it crashing onto the road—and onto anyone in the way. Besides the safety factor, I feel an obligation to avoid leaving rocks on the road or shoulder or destabilizing rocks that might end up there after I'm gone. Some excellent roadside localities have been officially closed because collectors failed to do this.

Be careful not to undermine trees that are clinging to the cliff or its top; they stabilize the whole structure in addition to providing shade and climbing aids for collectors and other creatures. The only time I really go nuts with the crowbar is in a roadside quarry off the road and intended for wanton destruction. When collecting near cliffs at beach localities, *never dig in the cliffs themselves*. Besides being extremely dangerous, this is also illegal.

Don't bite off more than you can chew. Some fossils are best left to be seen again or to be tackled by better equipped collectors. For example, if you find a whale skeleton in a cliff, contact a museum or a university. You'll never be able to unearth it and carry it off without damage to yourself, the fossil, or both, even if you do succeed in getting permission from the landowner to excavate it.

Sometimes a fossil is of appropriate size but the rock is too large to carry and too hard to break up. I once found a most unusual and beautiful brachiopod on the surface of a very large, very hard boulder. The rock resolutely defied my hammer and chisels. But my greed wouldn't let me give up, so I flailed away until at last the fossil itself was damaged. If you must have it, consider what special equipment is necessary, or consult an expert for assistance, or collect it with a camera.

One last thing: if I turn over a rock and find an ant colony or bunch of sow bugs, I figure that they got there first, and I try to replace the rock as I found it—especially in winter, when disturbance would kill them. It seems hypocritical to treat the remains of dead animals carefully and then casually devastate their living relatives.

Flies in the Ointment: Hazards in the Field

There are a few hardships in fossil collecting. Besides the physical exertion in climbing over rock exposures, extremes of heat and cold and a host of pests can drive you back into the modern world. Here is a list of some of the things I've encountered besides fossils—some welcome, others not: dogs, ticks, mice, lizards, snakes, wasps, dead animals, deer antlers, money, poison ivy, thistles and thorns, spiders and flies, butterflies, bird's nests, mosquitoes, crystals, arrowheads, and lots of broken glass. Of all the hazards, the last one is the worst. Some people cannot resist the thrill of throwing an empty bottle from a passing car and listening for the distant tinkling as it shatters on the rocks. So never hunt for fossils at a road cut with bare feet or thin-soled shoes.

Insects can be a major nuisance in warm weather, so bring along some repellant. Mosquitoes, especially, find fossil hunters to their liking, as they tend to be slow moving and too preoccupied to swat.

Beach collecting has its own problems. Some of the best vertebrate fossil collecting occurs in places where layers of clay are exposed. These sites can be treacherous, since wet clay is very slippery. Cleats might be worth getting if you intend to frequent such areas. Otherwise, go slowly and take little steps.

Other hazards that you might encounter at the beach include the familiar broken bottles and rusty cans and; in the warmer months, stinging nettles. Keep an eye open for patches of soft sticky mud. I've had some difficulty pulling myself out of ankle-deep muck more than once. Always avoid walking too close to unstable cliffs and keep alert for the sound of falling debris, particularly in the spring, when melting ice loosens bits, pieces, and the occasional slab of clay.

I've mentioned the fact that winter beaches are best for sharks' tooth collecting. Late fall, early spring, and mild winter days are usually best for inland collecting as well, because insects and undergrowth are at a minimum and the maximum amount of rock is accessible (assuming no snow is on the ground).

Many road cuts have been planted with ground cover to retard erosion and because some people actually find bare rock cliffs unattractive! In some places this growth becomes so heavy by midsummer that collecting is virtually excluded. Rock outcrops can also be dangerously hot in the full summer sun, so do your hot weather collecting in shorter stretches, or on cooler days, or in the mornings and evenings. A hat to keep the sun off your head is recommended, as is a supply of cool liquid refreshment.

When collecting at road cuts that are close to the highway, turn your face away when trucks pass by; they often leave a cloud of dust and flying pebbles in their wakes. One thing to get used to: passersby love to honk and shout at fossil hunters. I have no idea why.

Back Home Again

When a rock breaks to reveal a fossil within, it almost never exposes it completely. Bits of rock usually cling to the margins of the specimen or obscure large parts of it, so you'll want to clear away the unwanted rock before putting your fossil on display.

There are two rules of collecting that I learned the hard way. First, never give in to the urge to prepare your fossils in the field. I know, you don't want to take home a ten-pound chunk of rock for a little fossil in the left-hand corner. So you pretend to be a diamond cutter, and crack the little fossil instead of the ten-pound chunk. Or a trilobite's eye flies off into a pile of rock fragments. Don't bother to hunt—you'll never find it; at least, I never did. The second rule is patience. I have ruined in five minutes many gems that were protected by Mother Nature for millions of years before I got to them. Take time to plan a strategy for removing the matrix (the rock that surrounds your fossil), and resist the temptation to close your eyes, swing your hammer, and hope for the best.

I suggest working to free your fossil not only at home—and with patience—but while it is still embedded in a good-sized chunk of rock. The weight will give stability and strength, and your chisel or brush work will be less likely to shatter the rock or fossil. The same purpose is served by placing the rock on a small sandbag, which will conform to the contours of the specimen and help keep it together.

Have on hand an assortment of sharp instruments—pins, needles, jeweler's screwdrivers, small chisels and awls—to remove unwanted bits of matrix. A small hammer is necessary for using the small chisels, but tap lightly when working close to the fossil. It is much safer to remove the extra rock a chip at a time than all at once. I understand the best tools for this kind of work are in a dentist's kit. These kits can be purchased from surplus and catalog stores, or you can ask your dentist for his old ones.

When chipping matrix away from the borders of a fossil, avoid directing the force toward the specimen. Study the pattern of tiny cracks and layers in the rock and exploit them—or avoid them if they run through your fossil. Often the best way to attack a stubborn matrix around a fossil is to gouge out a trough around its margin, an eighth or a quarter of an inch from its edge. (Be sure you know where the edge is!) Then, gently use a small chisel to lever up the chips of stone between the trough and the edge of the fossil. Don't make the trough too deep, or you will lever up chips of your fossil as well.

Elmer's Glue-all is an even greater friend at home than in the field. It will compensate for most of the mistakes you make in judging the rocks. An old toothbrush is good for cleaning dirt and some kinds of mineral deposits from hard fossils, and small, soft-bristled paint brushes work well for delicate ones.

Some fossils are covered by a thin film of matrix that adheres to the surface of the specimen, obscuring the shine or detail. A needle or other scraping tool can be used to remove the thicker portions of this layer. If the shell material is harder than brass, a brass-bristled brush can be used to remove what remains. Attachments of brass-bristled brushes are also available for hand-held rotary power tools. Test the brush on an expendable fossil of the same material before trying it on a good one. If a hand lens does not reveal any abrasion of the fossil's surface—if the brass comes off on the fossil and not vice versa—then you're in business.

Don't press too hard with the brush; you might damage the fossil or the surrounding rock, and the buildup of brass on the surface may become too deeply embedded to come off easily. Soap and water and a toothbrush should remove the brass after you're through, though I've found this can be a slow process.

Some collectors spray their fossils with an acrylic coating, sold as artist's fixative. Spraying affords some protection to fragile fossils with the tendency to flake off, such as plant or fish fossils consisting of thin carbon films. It also strengthens the surface of extremely soft shales and sandstones that are shedding dust or fragments. Spraying can also improve the appearance of some specimens by increasing the contrast between the fossil and the matrix, giving the appearance of perpetual wetness, but it can also obscure differences in their textures that would make the fossil stand out. This difficulty may be circumvented by covering the matrix before spraying with pieces of paper cut to size or with masking tape, if the surface of the specimen isn't damaged by it.

I try to avoid spraying unless I see deterioration occurring; I don't like the shiny, unnatural appearance that spraying often results in. Some books say flat out to spray all fossils that consist of carbon films, but I have many plant fossils of this type that have held up for a decade or two without protection. As with most techniques, try it on an expendable piece and see what you think.

Eventually, you'll want to trim the rock to a reasonable size for your collection, which brings me to the first real expense to consider—a rock saw. Using a rock saw is by far the safest and most satisfactory way to remove large pieces of matrix that are not too close to the fossil. Most of the fossil-bearing rocks illustrated in this book have been trimmed in this way. My rock saw, which has a six-inch, water-cooled blade, a powerful motor, and overhead lamp, has put in ten years of good service with only two or three blade replacements. Comparable saws are available today for around $150 to $200. This tool will keep your rock tonnage to a minimum. By the way, these saws often come equipped with diamond-studded blades, which are capable of cutting very hard minerals and gems. Much less expensive blades are available, which are adequate for cutting most of the sedimentary rocks that fossils occur in.

After they are trimmed with the rock saw, specimens are often left caked with mud. A toothbrush can be used to clean up hard rocks and fossils, but use a soft-bristled brush to clean away the residue from soft or delicate specimens.

Some rocks can be successfully trimmed by hand (without a rock saw), particularly soft, thin-bedded shales. Try gouging out the outline where you want the rock to break with a chisel or awl. When this mark is at least half the thickness of the slab, place the specimen on the edge of a table, with the unwanted portion suspended. Hit this part sharply with a small hammer or the end of a chisel. And have the Elmer's Glue-all handy.

I have mentioned that silicified fossils in limestones may be dissolved out of the rock by acidic rain. But you don't have to wait for the hundreds of rains necessary to accomplish this. Take home a few hunks of rock whose surfaces show a silicified fossil content and immerse a sample in a dilute solution of muriatic acid (HCl) in a glass or plastic bowl (not metal), and the limestone will fizz away, leaving only mud and sand—and fossils. I have obtained some beautiful corals from Devonian limestones in this way. It is best to try a sample first, because some fossils are only partially silicified and may be partly dissolved by an acid solution.

As you may expect, the above technique involves several safety measures. When diluting the muriatic acid (which is available at most hardware stores), observe the AAA rule—*Always Add Acid to* water. This promotes better mixing, since the acid is heavier than water. More importantly, it lessens the danger of overheating, boiling over, and spattering, which may result from adding the water to the acid. Use cold water to further minimize this risk.

Add the acid until the rock begins to bubble vigorously, and continue adding at intervals to keep it bubbling. Muriatic acid gives off dangerous fumes, so do your pouring outdoors and in a safe place. The fizzing will send a fine spray of acid around the bowl area, so leave it in a place that is not accessible to small children, animals, or other unsuspecting visitors. The used acid should be disposed of carefully, as it may not have been completely neutralized by reaction with the limestone. Add powdered baking soda to the liquid until there is absolutely no sign of fizzing. This will complete the neutralization process, and you can then safely pour the solution down the drain while running the faucet to dilute the solution and flush the drain thoroughly.

After the rock has been dissolved, pick up the large fossils, rinse them thoroughly, and add them to your collection. Smaller fossils require more attention. The muddy residue at the bottom of the bowl should be carefully rinsed with water to remove the acid and then dried on a paper towel. A watercolorist's paintbrush is excellent for teasing out especially small fossils. Keep the brush wet, and the specimens will adhere to it. They can be cleaned of the mud that coats them by immersion in water.

Some fossils are difficult to collect and preserve because they occur in weak or poorly consolidated rocks. There are techniques for hardening such rocks, some involving the use of dangerous chemicals such as toluene and acetone. A safer treatment that has worked well for me is to soak the specimens in a solution of the fossil collector's friend—Elmer's Glue-All.

Dilute enough glue with an equal volume of water to cover the specimen in a small bowl. Leave it immersed for five or ten minutes so the liquid will soak all the way into the rock. (Shorter immersions may be necessary if the rock tends to disintegrate when wet.) Remove the specimen and dab it gently with an absorbent paper towel to remove excess solution; this will minimize shiny areas when the glue is dry. Remember that the rock is weaker and softer when wet, so don't wipe away the excess liquid or you may wipe away the fossil as well. Allow several hours for the glue to dry thoroughly. This technique is especially effective for consolidating porous or powdery rocks.

It is essential to keep a written guide to your collection, with a page or two for each locality. Give each collecting site a name, record directions in case you lose your field notebook, identify the formation or formations and their geologic periods, and make a list of all the fossils you found there, even the ones you didn't keep. (You'll soon begin to notice associations of fossils, which will help you recognize the same formations in the field.) This guide can be as simple or as elaborate as you want; it might even become a work of art.

For other techniques and a further discussion of the ones mentioned here, get a copy of *Fossils for Amateurs*, by Russell P. MacFall and Jay C. Wollin (Van Nostrand Rheinhold, 1983). It is the ultimate handbook for fossil preparation by hobbyists.

Where Do I Put Them?

As long as your collection continues to grow, where to house it is a never-ending challenge. I have enlisted every container imaginable to accommodate my collection and am convinced that there is no right way. But if you don't tackle the problem early, it won't take long for a fascinating collection to turn into "that mess in the garage." Actually, you'll probably want to start two collections: one for show pieces and one for second stringers. The latter category will include duplicates as well as specimens that are not well preserved but are valuable because they show variety or certain features better than your prize fossils.

Boxes and jars are fine for storing the extras, but you'll probably want to find a drawer or glass-covered case for the display specimens. Try to find or make display areas that are just deeper than the largest fossil is tall. Nothing wastes more space than a chest with a thin layer of fossils at the bottom of each drawer, and space will mean more than you might think at first.

It's a good idea to line the bottom of a display area with felt to discourage your specimens from rolling around, especially if they are in a drawer that is opened and closed frequently. Cut a piece of corrugated cardboard to fit, then cover it with inexpensive felt (available at fabric stores), and tape the overlapping material to the underside of the cardboard. Be sure to choose a color of felt that will contrast favorably with the colors of most of the specimens you plan to display. Open air collections are not recommended. There is nothing like a thick layer of dust to make your collection look uninteresting, and most fossils are too fragile to withstand frequent cleaning. Cover your display trays with sheets of double-strength glass, protecting the sharp edges with heavy plastic or cloth tape.

As soon as you have two hunting localities, you have a problem: how to tell their fossils apart. Many collectors number their specimens, applying a dab of white paint on the back of the rock and writing an index number on it. This number should correspond to a reference system devised for your collection guidebook, a card file, or a computer list. But what about sharks' teeth, whose back sides look as good as their fronts, or tiny fossils? If you decide to keep each specimen in a small, separate tray, there is no problem: number the tray or label it with all collecting information. This is the customary way to store and display fossils. I prefer to arrange the specimens together on a felt surface, segregated by locality (as in the illustrations in part 3)—and hope I don't spill the drawer.

I used to display my fossils segregated by type—all trilobites together, et cetera. The problem was that I had more difficulty getting a feel for the separate localities. Specimens from a locality usually have a common color or standard of preservation that you become attuned to more readily when you display them together, making it easier to recover from a spilled drawer. More important, displaying fossils from the same locality together helps you to get a sense of what that ancient community was like. The problem with this system is that not all the specimens from one locality are compatible in size. A fifteen-pound whale vertebra shouldn't share a drawer with tiny teeth and shells. So you might want to start a third and even a fourth collection: one for large specimens and one for tiny specimens.

If you plan to collect on a fairly large scale and if you have the space, a map file cabinet probably best accommodates average-sized fossils. These cabinets are heavy and expensive, but their numerous shallow drawers are ideal for most fossils. If you're handy with a hammer and saw, you might consider constructing such cabinets or display boxes custom-made for your collection.

I close this section with the confession that, despite an assortment of shelves, filing cabinets, boxes, jars, and drawers, my collection is never completely in order nor properly housed. You may want to tailor the size of your collection to the available space by keeping only the finest specimens. And yes, I've tried that, too.

Part III A Portfolio of Fossil-Collecting Localities

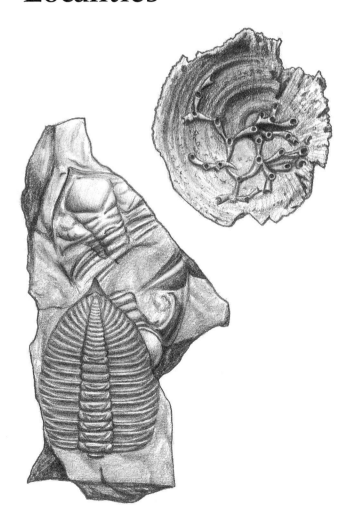

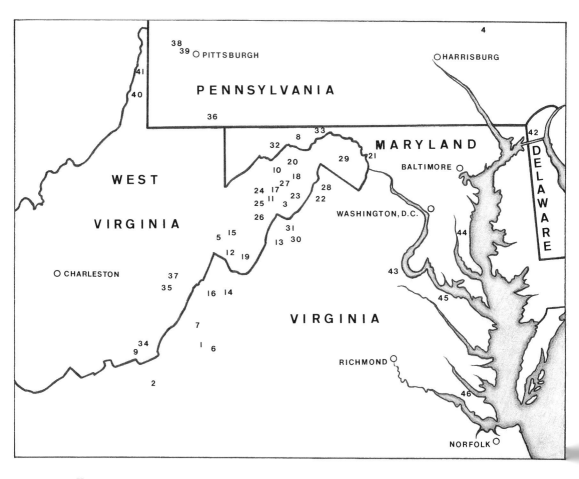

Collecting Localities in the Mid-Atlantic States (see Table 2)

There is much more to fossil collecting than collecting fossils. Many exciting and unexpected experiences with nature await you. In addition, you will derive benefits from the exercise and fresh air and from the mental quiet that settles in when you're engrossed in the search. This chapter shares some of the experiences I have had and gives specific information about my favorite localities. These are listed in table 2 and located on the map of the region. Some of these spots I learned about from other publications or from other collectors, but the majority of them I just happened on. With a few exceptions, these are not unique or "hot" localities; many equally good or better hunting grounds await discovery.

A word of caution: don't let your expectations run too high. So many times, on reading the particulars in a guidebook about some exposure, I arrive expecting to fill box after box with spectacular fossils. More often than not, I find little at first and become disappointed. Then I begin to find a few specimens, more modest in size than I had pictured, and then finally realize that I am doing pretty well after all. The fossils illustrated with each locality portrait represent a fairly typical collection for that locality. My experience suggests that you'll top at least one or two of these specimens each time you visit.

The fossil list for each locality gives genus and species for each fossil. In cases where the genus is known but the species is uncertain, the abbreviation sp. follows the genus name, meaning "a species of the genus given." For example, *Platyceras* sp. means the fossil is a species (unidentified) of the genus *Platyceras*.

It is worth remembering that paleontologists are continually reevaluating the classification of fossils and renaming genera and species. It often takes some time for these new names to work their way from technical journals into popular guidebooks. In fact, because many of the best technical references for identification are out of date, I have indicated the former names of some of the genera in parentheses after the new names (e.g., *Desquamatia [Atrypa] reticularis*).

The identification of some corals and many bryozoans to genus or species level is often impossible without studying the intricate details of their internal structures. Frequently, these details are not well preserved or can be seen only by cutting open, and therefore destroying, the specimen. For this reason, I give scientific names only for those forms that are particularly distinctive and easily recognized. Other types are given general descriptions, such as "branching species of bryozoan" or "horn coral."

Before I get started, I ask a favor: Please be gentle with these localities. I love them all, and you might some day as well; there's magic where the ages mingle.

For the sake of future collectors, and future visits of our own, let's try to keep these localities open and attractive to collectors.

I was digging up trilobites when an old man stopped and said, "Now if you're really interested in fossils, and want to see some special things, come with me." And something in his eyes said he knew something I needed to know. So I followed, thinking, Maybe this is the mountain man who knows the secret ways of fossil hunting: A master of the rocks.

He showed me his collection, his nondescript slabs: a few poor brachiopods; nothing I'd lean over for. But his eyes danced and his face shone with pride and wonder as he patiently explained each piece as he understood it—which was not at all. The only part of his collection I'd bother to keep was the trilobite I gave him. But I learned from his unjaded eyes that every stone and broken shell is a miracle. This old man of the mountain had learned the secret after all.

Table 2. **Collecting Localities in the Middle Atlantic States**

Locality	Period or epoch	Group or Formation	County and State
1	Ordovician	Lincolnshire	Rockbridge, Va
2	Ordovician	Liberty Hall	Montgomery, Va.
3	Ordovician	Martinsburg	Hardy, W. Va.
4	Ordovician	Martinsburg	Lebanon, Pa.
5	Ordovician	Reedsville	Pendleton, W. Va.
6	Ordovician	Reedsville	Rockbridge, Va.
7	Ordovician	Reedsville	Bath, Va.
8	Silurian	Rochester, McKenzie	Allegany, Md.
9	Silurian	Rose Hill	Monroe, W. Va.
10	Silurian-Devonian	Tonoloway, Helderberg	Hampshire, W. Va.
11	Silurian-Devonian	Keyser	Hardy, W. Va.
12	Silurian-Devonian	Keyser	Pendleton, W. Va.
13	Devonian	New Creek	Rockingham, Va.
14	Devonian	Corriganville	Highland, Va.
15	Devonian	Corriganville, Oriskany	Pendleton, W. Va.
16	Devonian	Licking Creek	Highland, Va.
17	Devonian	Needmore	Hardy, W. Va.

Table 2—*Continued*

Locality	Period or epoch	Group or Formation	County and State
18	Devonian	Needmore	Hampshire, W. Va.
19	Devonian	Needmore	Pendleton, W. Va.
20	Devonian	Needmore	Hampshire, W. Va.
21	Devonian	Marcellus	Washington, Md.
22	Devonian	Mahantango	Frederick, Va.
23	Devonian	Mahantango	Hardy, W. Va.
24	Devonian	Mahantango	Hardy, W. Va.
25	Devonian	Mahantango	Hardy, W. Va.
26	Devonian	Mahantango	Hardy, W. Va.
27	Devonian	Mahantango	Hampshire, W. Va.
28	Devonian	Mahantango	Frederick, Va.
29	Devonian	Mahantango	Berkeley, W. Va.
30	Devonian	Mahantango	Shenandoah, Va.
31	Devonian	Mahantango	Shenandoah, Va.
32	Devonian	Mahantango	Allegany, Md.
33	Devonian	Chemung	Allegany, Md.
34	Mississippian	Greenbrier	Monroe, W. Va.
35	Mississippian	Greenbrier	Pocahontas, W. Va.
36	Mississippian	Mauch Chunk	Fayette, Pa.
37	Pennsylvanian	Kanawha	Pocahontas, W. Va.
38	Pennsylvanian	Mahoning, Brush Creek	Beaver, Pa.
39	Pennsylvanian	Brush Creek	Allegheny, Pa.
40	Pennsylvanian	Monongahela	Ohio, W. Va.
41	Pennsylvanian	Monongahela	Brooke, W. Va.
42	Cretaceous	Mount Laurel	New Castle, Del.
43	Paleocene	Aquia	Stafford; King George, Va.
44	Miocene	Calvert, Choptank, and St. Mary's	Calvert, Md.
45	Miocene	Calvert, Choptank, St. Mary's and Eastover	Westmoreland, Va.
46	Pliocene	Yorktown	York, Va.

Locality 1 — near Effinger, Rockbridge County, Virginia

Lincolnshire formation, middle Ordovician period

Locality 1 is a large road cut overlooking Buffalo Creek on the southwest side of Route 251, just east of the village of Effinger, and 7.6 miles southwest of the junction in Lexington of Route 251 with U.S. Route 11. The exposure begins just southeast of the intersection with Route 612. The best collecting is at the gentle bend in the road about 200 yards southeast of Route 612. The shoulder is just wide enough for parking southeast of the guardrail near Route 612. More extensive parking is available in Effinger.

The dark gray limestones at locality 1 contain an abundance of bryozoans and other fossils. They are easily collected, as they are scattered loose on top of the reddish soil that covers much of the road cut. These fossils are silicified and have been etched out of the soluble rocks by acidic rainwater.

In addition to several species of bryozoans, brachiopods are fairly common, although most specimens are broken or somewhat distorted in shape. This distortion is due to the folding, stretching, and compression of the rocks and their contents during mountain building. Fragments of trilobites and gastropods turn up occasionally, as well as crinoid stem sections. The latter are common on the surfaces of limestone slabs near the junction with Route 612.

Notice the layers and nodules of black chert that occur in the limestone. These, and abundant silicified bryozoans, are characteristic of the Lincolnshire formation, which was formerly called the Lenoir limestone in this area.

In order to fully exploit this locality, you'll have to get down on your hands and knees and clamber over the roadcut, keeping a sharp eye on the surface for fossils. The exposure is steep and littered with loose rocks, so choose your footing carefully.

My own collection includes the following fossils from locality 1:

Brachiopods: *Camarella* sp.; *Hesperorthis* sp.; *Mimella* sp.; *Multicostella platys*
Bryozoans: branching, encrusting, and massive colonies
Crinoids: stems and segments
Gastropods: internal molds of unidentified species
Trilobites: *Homotelus* sp. (fragments); unidentified species (fragments)

Locality 1. *1, 2,* bryozoan colonies; *3, 4a, 4b,* brachiopods *Hesperorthis* sp.; *5,* brachiopod *Camarella* sp.; *6a, 6b, 7,* brachiopods *Mimella* sp.; *8,* brachiopod *Multicostella platys; 9–13,* bryozoan colonies. (Fossils shown 1.2x actual size.)

Locality 2 near Lusters Gate, Montgomery County, Virginia

Liberty Hall formation, middle Ordovician period

Locality 2 is a series of small road cuts on the north side of Route 785, 0.5 of a mile east of the village of Lusters Gate. To reach this locality from Blacksburg, turn east from Business Route 460 on Roanoke Street (Route 785) and continue 4.3 miles. The locality is on your left. The grassy shoulder across the road from the easternmost (furthest from town) exposure is the only parking available here—and is barely adequate. Traffic is usually light.

These beds of the Liberty Hall formation were formerly called the Athens shale. This locality is fairly well known by collectors and is a favorite haunt of geology students from Virginia Polytechnic Institute in Blacksburg.

The gray shales, which weather to tan, are sparsely fossiliferous and contain very few kinds of fossils, but most of the specimens that do turn up are well preserved, usually complete examples of the trilobite *Ampyxina scarabeus*. These trilobites are small, seldom exceeding three-quarters of an inch in length; I found one complete specimen less than one-eighth of an inch long. Slightly larger but a bit less common is the trilobite *Dionide holdeni*. The only other fossils present in any abundance are graptolites, some remarkably well preserved.

Most of my success was in the weathered, tan shales, but friends of mine collected many specimens from fresher material, in which the fossils were coated with an orange iron oxide residue, making them contrast nicely with the dark gray matrix. The following are the fossils I have collected at locality 2:

Graptolites: *Climacograptus* sp.
Trilobites: *Ampyxina scarabeus; Dionide holdeni*

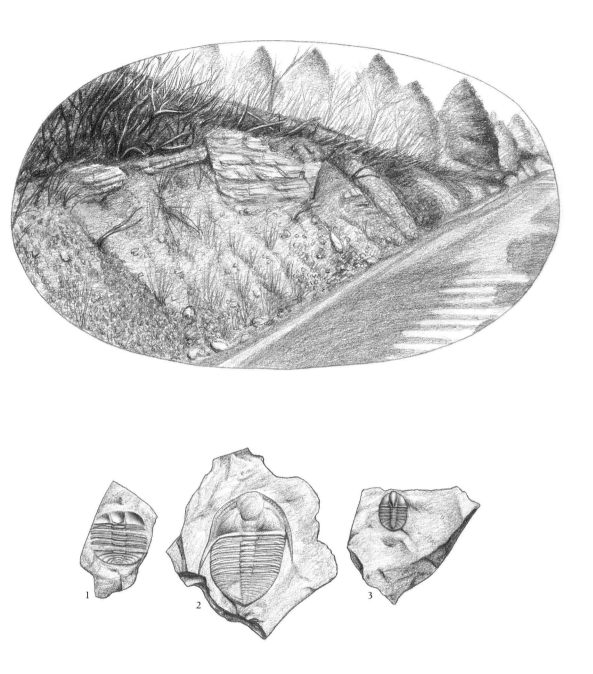

Locality 2. *1, 3,* trilobites *Ampyxina scarabeus* (the front of the head, or cephalon, is missing from *1*); *2, Dionide holdeni.* (Fossils shown 1.8x actual size.)

Locality 3 — near Perry, Hardy County, West Virginia

Martinsburg formation, late Ordovician period

Locality 3 is a small road cut on the east side of Route 23-10 at a sharp bend in the road, 0.6 of a mile north of the Wolf Gap recreation area (and the Virginia State line) and approximately 3 miles south of the village of Perry, West Virginia. (Route 23-10 is designated Route 675 in Virginia.) In addition to the road cut itself, the rocks that litter the hillside immediately across the road contain fossils. There is just enough shoulder across the road from the outcrop for parking. Traffic is very light.

The dark, green, gray, and brown shales exposed in this road cut have been distorted and cracked by the pressures that built the Appalachian Mountains, and most of the surface rocks have been reduced to chips and slivers by weathering. The fossils they contain are small, difficult to collect, and frequently twisted, stretched, or squashed. In fact, if not for a lucky find in the first few seconds of my initial visit, I would probably have decided it wasn't worth my time to collect here—and would never have come to know one of my favorite localities.

That first fossil was a partially enrolled trilobite (*Flexicalymene*), and subsequent searching has made me realize how lucky I was to find it, for this type of fossil is not very common here. Much more abundant are the fringed heads of the blind trilobite, *Cryptolithus*. Very occasionally, whole specimens of this fossil turn up, though an undistorted example is rare. These are some of the most elegant, gracefully designed trilobites I've ever seen. The fringes are perforated by numerous tiny holes, which give them the appearance of screens or filters. Suggested explanations for the purpose of these fringes include stabilizers for soft mud (like a snowshoe), filter-feeding mechanisms, or sensors for orientation in underwater currents.

This is a very challenging locality. Patience and careful collecting are rewarded more than energetic excavation. Take the time to look carefully over the surface of the layers and the rock chips. In most cases, the collectible specimens are the molds and impressions left behind by the weathering away of the calcium carbonate shell. These ghosts are often yellow or orange as the result of iron oxide residues. Since most of the shale here is dark green, these colored fossils are very pleasing to the eye.

Some layers are packed with crinoid segments and, very occasionally, complete crinoids, consisting of the long stems with the crowns, or calices (singular, calyx) still attached. Look for signs of these layers on the sides of large chunks of

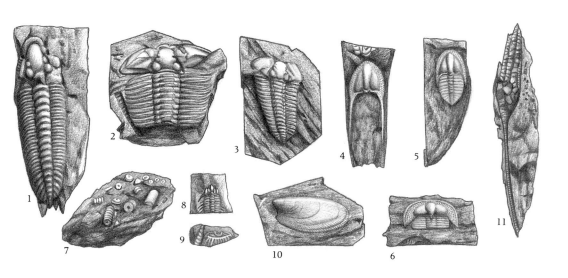

Locality 3. *1–3,* trilobites *Flexicalymene granulosa* (the differences in the proportions of these specimens are due to the deformation of the rocks); *4–6,* trilobites *Cryptolithus tesselatus* (the long spines seen on *4* are missing from the other two specimens); *7,* crinoid stem fragments; *8, 9,* trilobites *Odontopleura* sp. (*9* is an isolated tail, or pygidium); *10,* pelecypod *Ambonychia* sp.; *11,* crinoid *Ectenocrinus simplex* (stem with crown, or calyx, attached). (Fossils shown .9x actual size.)

shale and split them to expose the fossils. The best preserved specimens of all types are usually in less fossiliferous rock just above or below these crowded layers.

When collecting at this locality in the warmer seasons, watch for the large ants that prosper here. They may turn nasty if you disturb a nursery or other population center under a boulder or in a pile of rock chips. Other residents include ring-necked snakes, averaging about twelve inches in length. I nearly always find a couple or more during the course of my warm-weather collecting. They are harmless; in fact, I have handled scores of them and have never known one to even attempt to bite, though some will discharge a smelly substance from their cloacas when disturbed. (Actually, the odor isn't particularly objectionable.) If you're not afraid of snakes, take a moment to visit with one. As they are very vulnerable to your activities, living between and beneath the rocks, be careful to avoid crushing them.

This locality is a good place to settle down in the shade with a hammer and a magnifying glass and pick over a small area and let your mind drift back in time to a muddy sea where the trilobites were as thick as flies. The fossils I have collected here are:

Brachiopods: *Onniella (Dalmanella) fertilis; Pseudolingula* sp. (similar to *Lingula,* somewhat rectangular); *Sowerbyella* sp.
Bryozoans: branching, encrusting, and massive species
Cephalopods: *Michelinoceras* sp.
Crinoids: *Ectenocrinus simplex* (complete animals) stems and segments
Graptolites: *Climacograptus* sp.
Pelecypods: *Ambonychia (Byssonychia)* sp.
Trilobites: *Calliops* sp.; *Cryptolithus tesselatus; Flexicalymene granulosa; Isotelus* sp.; *Odontopleura (Acidaspis)* sp.; *Triarthus* sp.

Locality 4 at Swatara Gap, Lebanon County, Pennsylvania

Martinsburg formation, late Ordovician period

Locality 4 is a roadside quarry immediately underneath Interstate Route 81, on the west side of Route 140, 1.9 miles north of the town of Lickdale. To reach this quarry from Route 81, take exit 30 (Route 72 to Lebanon), follow Route 343 east to the stop sign in Lickdale, and turn left (north) on Route 140 (to Route 443). The locality is approximately 21 miles north of Harrisburg. There is a parking area in the quarry. *Be careful entering or leaving this locality, as visibility is very restricted, especially for southbound motorists on Route 140.*

Swatara Gap is one of the best known sources of trilobite fossils in the eastern United States. I had seen photographs of specimens from this locality since childhood, and when I approached it for the first time I half expected to find dozens of collectors milling around, a concession stand or two, maybe even a motel, museum, and gift shop. So I was pleasantly surprised to find an empty quarry and even more pleasantly surprised to find it full of easily collected trilobite fossils.

This is the same formation and basically the same assemblage of fossil types as are found at locality 3. But while specimens at locality 3 are often distorted and riddled with cracks, Swatara Gap fossils are mostly well preserved and much easier to retrieve. The gray and green shales contrast attractively with the yellow to orange iron oxide that coats most of the fossils, a feature also in common with locality 3.

The most distinctive and abundant fossil here is the trilobite *Cryptolithus tesselatus*. Examples of this species from this locality are shown in many field guides to fossil collecting. In about three hours, my brother David and I found a couple of dozen partial specimens and six complete ones. Besides trilobite fossils, Swatara Gap is also known for its starfish specimens. Several species of this kind of fossil occur here, but we have failed to turn up any. We did, however, find some intricate fossils of partial crinoid calices.

Our best luck—a bit of a miracle—was in the bedrock and rockfall in the southeastern part of the quarry (to your left from the entrance). There, David found half of a complete *Cryptolithus* near the top of a huge mound of rock chips. He also found the external mold of a complete specimen nearby. A half hour or so later, at the bottom of this mound of debris (which was at least forty feet high); he found yet another half of a whole trilobite. He gave all three of these fossils to me. A shiver went up my spine when I realized that all three were the remains of the same trilobite; the two halves fit together perfectly (see illustration).

The same fossils occur at a large road cut on Route 81, directly above this small quarry. But collectors have brought so much rubble onto the shoulder and road below the exposure that the state police have closed this locality. It is fortunate that the lower quarry is still open to the public—for an occasional miracle to happen.

The following fossils from this locality are in my collection:

Brachiopods: *Onniella (Dalmanella)* sp.
Cephalopods: *Michelinoceras* sp.
Crinoids: stems, segments, partial calices
Trilobites: *Cryptolithus tesselatus; Flexicalymene* sp.

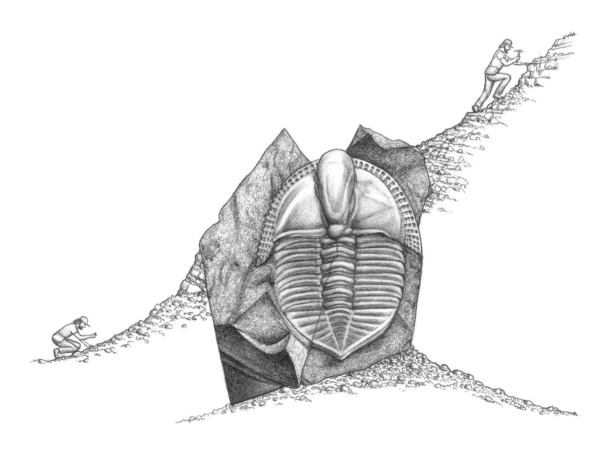

Locality 4. Trilobite *Cryptolithus tesselatus*. The trilobite itself is three-quarters of an inch long.

Locality 5
vicinity of Germany Valley Overlook, Pendleton County, West Virginia

Reedsville formation, late Ordovician period

Locality 5 is a series of large road cuts on the south side of Route 33, extending for approximately 0.5 of a mile east and 0.5 of a mile west of the Germany Valley Overlook (marked by a sign), 12 miles west of Franklin. The rocks are best exposed in the wide bends in the road as it climbs the steep side of North Fork Mountain.

Besides the parking provided at the overlook, the shoulders across the road from the outcrops are wide, sometimes extremely so. Trucks are reluctant to slow down for anything as they struggle with the gradient, so be alert, especially where the shoulder along the cliffs is narrow. Traffic can be moderately heavy.

Most of the rocks at this vast locality are fossiliferous, some layers being packed with crinoid stems or brachiopods or trilobite fragments. Most of the layers are gray, green, or brown shale, but significant amounts of sandstone are also present.

The upper layers of the Reedsville formation, which are predominately sandstones, are exposed in the cliff that juts into the road at its bend about a half mile east of the overlook. (This formation was once considered to be part of the Martinsburg formation and is frequently shown as such on older geologic maps.) This outcrop is shown in the illustration. About three feet above the road level (corresponding to the lower center of the picture) is a layer of calcareous sandstone, which is chock-full of brachiopods (*Orthorhynchula*) and pelecypods (*Ambonychia*). This is known as the *Orthorhynchula* zone and is found in the Reedsville formation over a wide geographic area. The specimens here are large and abundant and can frequently be extracted whole from the matrix.

Other exposures begin across the road from the overlook, at the next bend in the road (out of view and to the right in the illustration), and continue for a half mile or so to the west. The rocks here are mostly shales, though some fine-grained sandstone layers also occur. Both kinds of rocks may be extremely calcareous, with so much shell material that they are virtually made of fossils. A difficulty with these layers is that individual fossils are hard to extract. Also, they contain mostly fragmentary fossils, especially large smooth pieces of the trilobite *Isotelus*. Complete trilobite specimens are hard to come by and nearly impossible to retrieve intact—witness the mangled *Flexicalymene* specimen in the illustration.

As is so often the case in calcareous shales and sandstones, the best fossils are found in rocks from which the calcium carbonate has been removed by dissolution. These rocks are usually reddish brown here, as opposed to the grays and greens of fresher material. The shales often contain abundant crinoid stems and segments, while straight-shelled nautiloid fossils are common in the fine-grained sandstones. Both kinds of rocks contain numerous trilobite fragments.

The view from the overlook is spectacular and the locality is considered one of the most fossiliferous sites in the Reedsville formation. I have only begun to exploit it. The fossils I have collected are:

Brachiopods: *Lingula* sp.; *Onniella (Dalmanella)* sp.; *Orthorhynchula linneyi; Rafinesquina alternata; Zygospira* sp.
Bryozoans: branching, encrusting, and massive species
Cephalopods: at least three species of straight-shelled nautiloids, including *Michelinoceras* sp.
Crinoids: segments and stems
Gastropods: *Sinuites* sp.
Graptolites: *Climacograptus* sp.
Ostracodes: very large (⅜-inch-long) species
Pelecypods: *Ambonychia (Byssonychia)* sp.
Trilobites: *Calliops* sp.; *Ceraurus* sp.; *Flexicalymene granulosa; Isotelus* sp.

Locality 5 *(opposite).* 1, graptolites *Climacograptus* sp.; 2, 3, crinoid stem sections; 4, bryozoan colony; 5, 6, pelecypods *Ambonychia radiata;* 7, 8, nautiloids *Michelinoceras* sp. (note fragments of brachiopods *Onniella* sp. on 8); 9, 10, brachiopods *Orthorhynchula linneyi;* 11, trilobite *Flexicalymene granulosa;* 12, trilobite *Flexicalymene* (partial heads), fragments of brachiopods *Onniella* (fine ribbed) and *Zygospira* sp. (coarse-ribbed). (Fossils shown .7x actual size.)

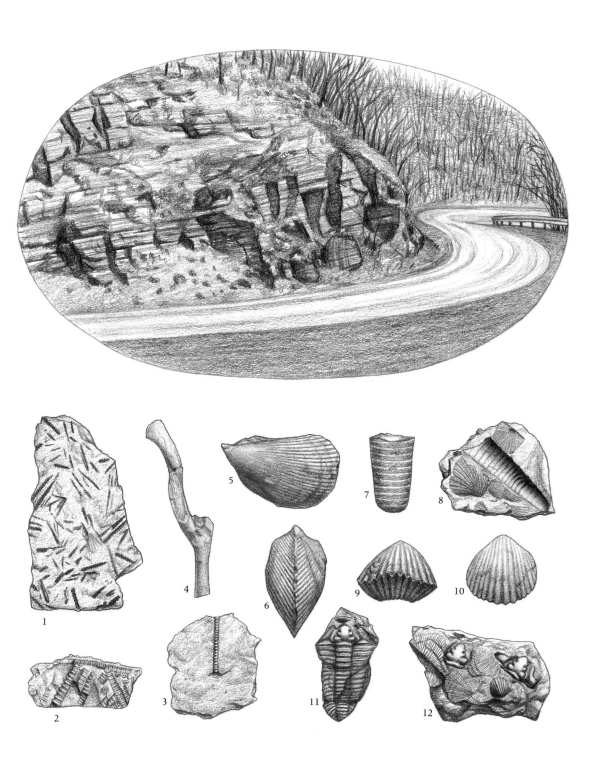

Locality 5, Germany Valley Overlook, West Virginia   55

Locality 6   near Lexington, Rockbridge County, Virginia

Reedsville formation, late Ordovician period

Locality 6 comprises extensive road cuts on the north side of U.S. Route 60, 1.4 miles east of the junction with U.S. Route 11 in Lexington and 0.8 of a mile west of the junction with Interstate 81. The exposures extend east and west of the entrance to a private drive on the north side of Route 60, 0.2 of a mile west of Route 699. The best collecting is in the exposures to the west of the private road.

Traffic on Route 60 is moderately heavy, but the shoulders are wide and the outcrops fairly well back from the road. Wide gravel areas near the entrance to the driveway and at the westernmost end of the locality afford the best parking.

A great thickness of Ordovician rocks is exposed in steeply tilted beds here. Most of the layers are of dark gray limestone and shale, tending to light gray in the eastern exposure. The best collecting seems to be in the darker layers in the western road cut, where lens-shaped colonies of the massive bryozoan *Prasopora* are abundant. Many of the dark limestone layers are full of brachiopods (*Rafinesquina* and *Onniella*), although complete, undeformed specimens are not common. The shale layers, which weather to green, brown, or light yellow, are less fossiliferous but do contain brachiopods; one yellow fragment I found bore the head of a trilobite (*Ceraurus*).

It would take a team of geologists quite some time to isolate and exploit all the fossiliferous layers exposed here, and I have only scratched the surface. A quick survey of the eastern road cut yielded some slabs of limestone, whose weathered yellow surfaces were littered with small, well-preserved bryozoans (*Rhinidictya*), brachiopods (*Onniella*, *Sowerbyella*), and pieces of trilobites (*Cryptolithus*). At this locality, successive "pages" of Earth's history are especially easy to read, since they are thin, accessible, and stacked on end.

Locality 6 *(opposite).* 1–5, bryozoan colonies *Prasopora simulatrix* (5 shows fragments of several specimens); 6, brachiopods *Sowerbyella rugosa*; 7, 8, brachiopods *Rafinesquina alternata*; 9, 10, brachiopods *Onniella fertilis*; 11, bryozoan colonies (specimens 1–11 shown .6x actual size); 12, bryozoans, crinoid segments, brachiopods, and a trilobite tail (*Calliops* sp.) at bottom center; 13, bryozoan *Rhinidictya nicholsoni*; 14, bryozoan; 15, trilobite *Ceraurus* sp. (partial cephalon). (Specimens 12–15 shown 1.2x actual size.)

Locality 6, Lexington, Virginia

I have collected the following here:

Brachiopods: *Onniella (Dalmanella) fertilis; Rafinesquina alternata; Rhynchotrema increbescens; Sowerbyella rugosa*
Bryozoans: *Prasopora simulatrix* (disc-shaped or hemispherical colonies); *Rhinidictya nicholsoni* (ribbonlike colonies); several branching species
Crinoids: stems and segments
Gastropods: fragments of internal molds
Trilobites: *Calliops* sp.; *Ceraurus* sp.; *Cryptolithus* sp.; *Isotelus* sp. (fragment of very large individual)

Locality 7 — near Warm Springs, Bath County, Virginia

Reedsville formation, late Ordovician period

Locality 7 is a road cut on the north side of Route 39, 1 mile west of the intersection with Route 220 in the town of Warm Springs, Virginia, and just east of Route 692.

The shoulder on the south side of the road across from the road cut is just wide enough for parking. Traffic is light.

The light gray shales and limestones exposed here contain a limited variety of fossils, but most specimens are very well preserved. Bryozoans are the most abundant, especially small, branching species and the large (up to three-inch diameter), cone-shaped colonies of *Prasopora*. The intricate structure of both may be seen with a hand lens.

Many of the slabs of limestone are covered with a thin, tan or light gray, shaley layer, which contains brachiopods, gastropods, and bryozoans. The limestone layers themselves are sometimes crowded with tiny, complete specimens of the brachiopod *Zygospira*, few of which exceed one-eighth of an inch in width. These brachiopods were among the earliest members of the order Spiriferida. Spiriferid brachiopods attained tremendous variety and abundance in the middle to late Paleozoic seas.

The fossils in these rocks have close affinities with those at locality 6 in similar limestone layers of the Reedsville formation. The shales and sandstones of the upper part of that formation may be seen in road cuts just west of locality 7. The difference in the character of the rocks reflects the increase in mountain-building activity as the Ordovician period progressed. These rocks are overlain by the red beds of the Juniata formation, which crop out about a quarter mile to the west.

Fossils from locality 7 in my collection are:

Brachiopods: *Onniella (Dalmanella) fertilis; Zygospira* sp.
Bryozoans: *Prasopora simulatrix;* branching species
Gastropods: *Liospira micula*

Locality 8 near Cumberland, Allegany County, Maryland

Rochester and McKenzie formations, middle Silurian period

Locality 8 is a large road cut on the north side of U.S. Route 40-48, 6 miles east of the city of Cumberland, Maryland. The exposure is located 1 mile west of the Pleasant Valley Road entrance to Rocky Gap Park and 4 miles east of the junction of Route 40-48 with U.S. Route 220 North. (Locality 33 is 11 miles east of this site.)

The shoulder at this locality is wide enough to permit parking, but traffic is fairly heavy and road construction in the area may complicate matters.

Both the Rochester and McKenzie formations are middle Silurian in age, and both consist of medium to dark gray shale and limestone. The McKenzie lies directly on top of the Rochester, and the only way to recognize the boundary between them is through a careful comparison of the fossil contents of specific layers. Because many of the fossils at this locality are found in loose rocks, and a few species occur in both formations, I consider them as a single unit here.

Generally speaking, the eastern portion of the road cut is best for thin layers of very fossiliferous limestone, often packed with the brachiopod *Cupularostrum* as well as a few bryozoans, large round masses of coral (*Favosites*), the extremely rare trilobite, and innumerable tiny black ostracodes. The western two-thirds of the exposure also has its share of ostracodes, as well as numerous large fossils of the straight-shelled nautiloid *Michelinoceras*, some nearly a foot long and two inches in diameter and fragments of even larger ones.

Some of the nautiloid specimens have weathered in such a way that the internal structure is visible. In addition to the different chambers of the shell, the central tube that connects them, called the siphuncle, can be seen.

I am sure there are many more kinds of fossils here than I have found; it is a vast outcrop, and specimens have turned up in nearly all parts of it. However, with the exception of the nautiloids, most of the fossils blend into the surface texture and color of the rocks, so you will have to be observant to exploit this locality fully. The following are the fossils I have collected:

Brachiopods: *Craniops(?)* sp.; *Cupularostrum* sp. (very common); *Lingula* sp.
Bryozoans: branching species
Cephalopods: *Michelinoceras* sp.
Corals: *Favosites niagarensis*
Gastropods: *Murchisonia (Hormotoma) marylandica*
Ostracodes: *Kloedenella* sp.
Pelecypods: *Cleidophorus nitidus*
Trilobites: *Calymene* sp.

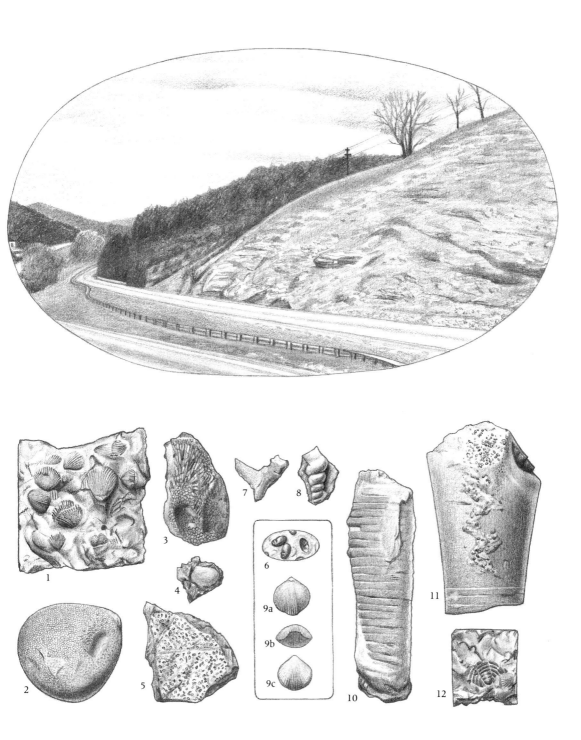

Locality 8. *1*, brachiopods *Cupularostrum* sp.; *2, 3* corals *Favosites niagarensis; 4*, pelecypod *Cleidophorus nitidus; 5, 6*, ostracodes *Kloedenella* sp. (the specimens in 6 are shown 4.3x actual size); *7*, bryozoan; *8*, gastropod *Murchisonia marylandica; 9a–9c*, brachiopods *Cupularostrum* sp. (shown actual size); *10, 11*, nautiloids *Michelinoceras* sp. (note ostracodes on *11*); *12*, trilobite *Calymene* sp. (tail). (Specimens other than 6 and *9a–9c* shown .6x actual size.)

Locality 9 near Waiteville, Monroe County, West Virginia

Rose Hill formation, middle Silurian period

Locality 9 is a small, weathered road cut on the north side of Route 635, 3.3 miles southwest of Waiteville, West Virginia, and 1.7 miles northeast of the Virginia State line. Fossils also occur in boulders on the hillside across the road from the outcrop. This road is lightly traveled and the shoulder opposite the road cut provides ample parking.

The Rose Hill formation (formerly known as the Clinton formation in this area) is known for its iron ore deposits. Fossils occur here in very soft sandstones, ranging in color from light gray to lavender to reddish purple. The water conditions in which these sediments accumulated must have been fairly turbulent, since all the trilobites I have found are disarticulated—isolated tails, heads, and thoracic segments, often piled on top of each other, and surrounded by hordes of tiny (one to two millimeters long) ostracode fossils.

While preservation of surface detail is only fair, the abundance of specimens and their attractiveness makes this a locality worth visiting. The fossils themselves are usually yellow, orange, or even pink and contrast strikingly with the different colors of matrix. The most abundant trilobite fossil is *Calymene cresapensis*. Less common and very similar in appearance is its relative *Liocalymene clintoni*. The tails of *Liocalymene* are roughly the same size and shape as those of the more common species, but their sides are smooth rather than ribbed, as in *Calymene*.

Many of the most fossiliferous boulders I found were on the hillside across the road from the exposure. Only about one rock in ten is productive, but those that are are usually packed with specimens. If you collect there, be careful of the trash that people have dumped; rusty cans and broken bottles turn up nearly as often as trilobites.

The fossils I have collected include:

Brachiopods: *Cupularostrum (Camarotoechia) neglectum;* unidentified species
Bryozoans: encrusting species
Cephalopods: straight-shelled nautiloids
Ostracodes: *Zygobolbina conradi*
Pelecypods: unidentified species
Tentaculitids: *Tentaculites* sp.
Trilobites: *Calymene cresapensis; Dalmanites* sp.; *Liocalymene clintoni;*
   Trimerus sp.

Locality 9. *1*, ostracodes *Zygobolbina conradi; 2–5*, trilobites *Calymene cresapensis* (*2* and *4* are partial heads, *3* and *5*, mostly tails); *6*, trilobite *Liocalymene clintoni* (tail); *7*, brachiopod (in center) *Cupularostrum neglectum* (1.2x actual size); *8*, ostracodes *Zygobolbina conradi* (2.4x actual size); *9*, brachiopod, internal mold of unidentified species; *10*, trilobite *Trimerus* sp. (partial tail); *11*, trilobite fragments *Calymene cresapensis*. (Fossils other than 7 and 8 shown .6x actual size.)

Locality 10 near Romney, Hampshire County, West Virginia

Tonoloway formation and Helderberg group, late Silurian and early Devonian periods

---

Locality 10 consists of the gravel and boulders on the eastern fringe of a gigantic commercial quarry on the north side of U.S. Route 50, 2.2 miles west of Romney, West Virginia. Though the quarry itself probably contains fossils, it is not open to collectors. (Better specimens can probably be found in the more weathered and accessible rock on its margins, anyway.) There are wide shoulders on both sides of the road. Be careful to park out of the way of traffic going in and out of the quarry.

This locality is a quick stop, with fairly limited collecting opportunities but of special interest because of the silicified corals that may be found here. On my first visit, I noticed some rather unimpressive-looking corals on the surfaces of large boulders on the hillside east of the quarry. I broke off a small chunk of this rock, put it in my third-string collection, and forgot about it. A few months later I came across this chunk and got the idea of dropping it in muriatic acid (see part 2, How to Collect Fossils, for a description of this technique). I was dumbfounded by the result: an elegantly branching white cluster of *Favosites* coral, contrasting with the remaining, very dark gray limestone.

On subsequent visits to locality 10, I collected much more of this material, including one ten-pound rock that proved to be almost made of fingers of coral, some as long as four inches. Other fossils, especially bryozoans, crowd the surfaces of some of the rocks littering the shoulder of the road and the hill in the area. The distinctive plates from crowns of the cystoid *Pseudocrinites* often occur on these specimens. Also found at this locality are slabs of thin-bedded limestone covered with fossils of the unusually large ostracode *Leperditia*. As the black shells of these fossils are weathered away, dark oval outlines remain.

All of these rocks and fossils come from the late Silurian Tonoloway formation or from the Helderberg group, which straddles the Silurian-Devonian boundary. The Helderberg group consists of a succession of formations, mostly limestone, including the Keyser formation, which is partly Silurian, and the Devonian New Creek, Corriganville, and Licking Creek formations. Any or all of these rock units may be exposed in the quarry.

Since the fossils mentioned here were all found in loose material, it is difficult to say which formation each specimen came from. Most of the fossils at this locality, including the silicified corals and the bryozoan-covered slabs, probably come from the Keyser formation, particulary those bearing the distinctive cystoid

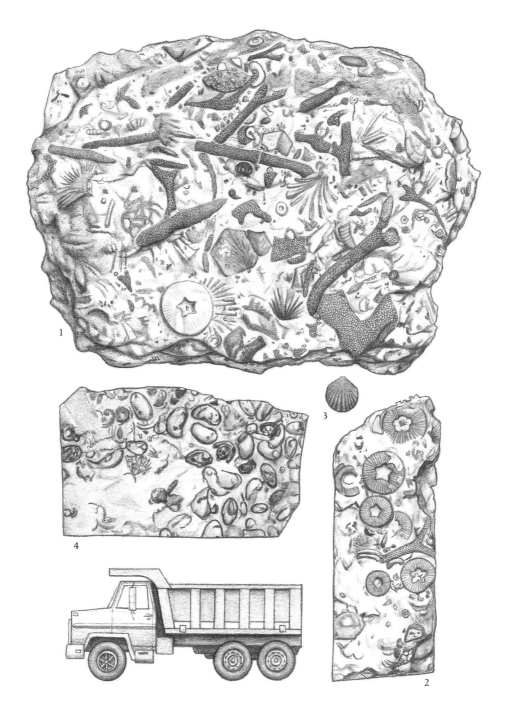

Locality 10. *1,* slab with bryozoans, brachiopods (including *Rhynchotreta* sp. just above and to the right of center), and cystoid segments and a plate from crown, or calyx, of *Pseudocrinites* sp. (just below center); *2,* bryozoan and cystoid segments; *3,* brachiopod *Cupularostrum litchfieldense; 4,* ostracodes *Leperditia* sp. (Tonoloway formation). (Fossils shown 1.7x actual size.)

plates. Slabs with the large *Leperditia* ostracodes are derived from the underlying Tonoloway formation. But in the absence of such guide fossils, it is best to be cautious in assigning specimens to specific formations.

Fossils in my collection include:

Brachiopods: *Cupularostrum (Camarotoechia) litchfieldense; Howellella vanuxemi; Rhynchotreta* sp.
Bryozoans: several branching species
Corals: *Favosites limitaris*
Cystoids: *Pseudocrinites* sp. (segments and plates from calices)
Ostracodes: *Leperditia* sp.

Locality 11 along the Lost River near Wardensville, Hardy County, West Virginia

Keyser formation, late Silurian to early Devonian periods

Locality 11 fossils are found in cliffs and gravel along the north shore of the Lost River, 5 miles west of the town of Wardensville, where it runs just south of Route 55-259. To reach these cliffs, park on the wide shoulder on the south side of the road (another very wide parking area is just west on the north side of the road), and follow the trails down to the river. The trail that goes to the left and follows a rocky streambed is best, as some of the trails that go straight ahead or to the right end abruptly at the top of the cliff. Watch for barbed wire near the bottom of the hill on the lefthand trail. Fishermen use this area quite often, so crossing the fence is apparently tolerated by the landowner.

The diver glides through the shallows, his shadow catching corals that jut from the bottom. The fish drift sideways, riding the current to new hiding places. He rises for a breath, slaps the water and dives again, a trail of bubbles wobbling up beside him, merging with the cold mountain air—for this lagoon is a freshwater pond on a chilly limestone stream. The frozen corals, their tiny homes filled with quartz, once lived in an antique sea, once huddled in the core of a mountain. And now the reef has been reborn, to decorate a beaver's world.

Unfortunately, the beaver seems to have disappeared, or has built a new lodge elsewhere along the river. The cliffs border a very deep pool, which you will get to know intimately if you don't choose your way carefully. As you head upstream (to the right from the trail), a series of nearly vertical layers of limestone and shale jut out into the river. Some of the softer, more soluble limestones in the recesses have abundant corals (*Favosites*) on their surfaces.

If you continue along the shore for fifty yards or so, you will come to a cove beyond which you can't proceed easily. (A cave is located above this cove, which I have not explored.) The rocks exposed just downstream of the cove—and those that jut out into the river here—have produced the best fossils for me.

Besides *Favosites*, the chain coral *Halysites* may also be found. The presence of *Halysites* in particular layers is considered an indication that they were laid down during the Silurian period, since that coral seems to have become extinct before the Devonian period began. Look for loose pieces of coral along the shore that have been dissolved completely free from the limestone. These are silicified and of the same type as those found at locality 10.

This is a picturesque spot, and though the fossils are not particularly varied (and I miss the beaver), I've found some nice specimens here:

Brachiopods: *Cupularostrum* sp.
Corals: *Favosites helderbergiae* (usually massive); *Favosites limitaris* (branching); *Halysites catenularia*
Crinoids: stems and segments
Stromatoporoids: unidentified species

Locality 11 *(opposite)*. *1,* coral *Favosites helderbergiae; 2,* coral *Favosites limitaris; 3,* coral *Halysites catenularia; 4,* crinoid stem section; *5,* beaver lodge.

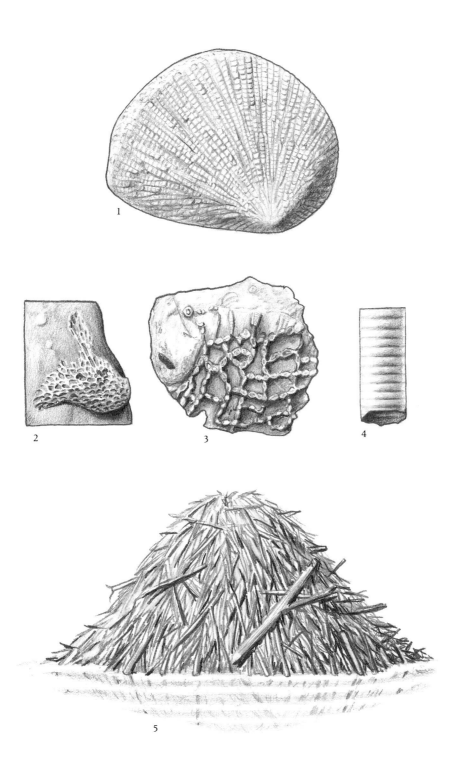

Locality 11, Lost River, West Virginia

Locality 12 along Thorn Creek near Franklin, Pendleton County, West Virginia

Keyser formation, late Silurian to early Devonian periods

Locality 12 fossils occur in road cuts and on the hillside along Route 23 (Johnston Road), which parallels Thorn Creek. From the junction of routes 33 East and 220 South in the town of Franklin, take Route 220 south for 3.3 miles, turn left (east) on Johnston Road (Route 23). Continue for 1.2 miles, crossing the South Branch of the Potomac River, turn right, go 0.1 of a mile and turn left, then proceed 5.75 miles to the locality. Roads are poorly marked; when in doubt, always follow the larger stream.

Thorn Creek is an excellent trout stream and is jealously guarded by private landowners, so legal parking areas are few and far between. Fortunately, a wide shoulder is located directly across the road from the outcrop. Traffic is very light.

The road along Thorn Creek is bordered by limestone cliffs for several miles in this area. I have explored only one small stretch of a half mile or so and found fossils throughout, so many more "hot" spots are probably nearby. Most of the specimens may be collected whole from the surface, having been weathered out of their limestone matrices. This makes for easy collecting and beautiful fossils. Though most specimens are fairly small, they are abundant, and preservation of detail is excellent.

It may take a while for you to begin to distinguish the fossils from the many odd chips of stone that also litter the hillside, but once you develop your search images, you should start picking them up right and left. Especially common are the small, smooth, almost clam-shaped brachiopods of the species *Nucleospira elegans*. Fossils may also be collected from the dark gray limestone boulders from which the loose examples are derived. But if you intend to use acid to retrieve individual specimens, run some experiments on expendable pieces first; my one effort in this line yielded unsatisfactory results, with loss of surface detail on the fossil.

Cystoid and bryozoan fossils are best represented on the surfaces of the flattened, thin slabs of shale that separate the thicker-bedded limestone layers and also litter the exposure. These rocks are light gray when fresh, weathering to tan or light yellow, and the dark gray or brown fossils often stand out in high relief and contrast on the surface. I found some cystoid fossils in this material consisting of fairly long, branched sections of stems or arms of the crowns. A complete crown preserved in this way would be spectacular. It is possible that some of the stems and segments that look like cystoids really come from crinoids, but as there is no

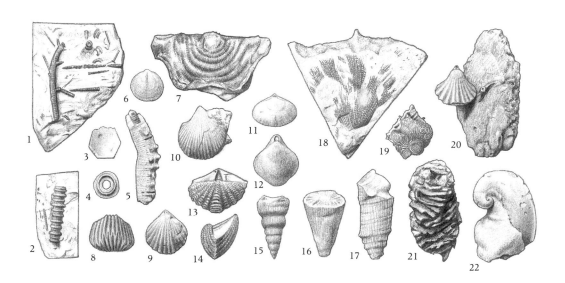

Locality 12. *1–5*, cystoid fragments; *6*, brachiopod *Rhipidomella* sp.; *7*, brachiopod *Leptaena* "*rhomboidalis*"; *8, 9*, brachiopods *Cupularostrum litchfieldense*; *10*, brachiopod *Uncinulus* sp.; *11*, brachiopod *Nucleospira elegans*; *12*, brachiopod *Meristella* sp.; *13, 14*, brachiopods *Cyrtina dalmani*; *15, 16*, corals *Enterolasma* sp.; *17*, poriferan, or sponge; *18*, bryozoan *Fenestella*; *19*, calcareous algae; *20*, brachiopod *Rhynchotreta* sp.; *21*, stromatoporoid; *22*, gastropod *Platyceras* sp. (Fossils shown .8x actual size.)

easy way to tell such fragments apart—and because I know cystoids are present—I've assumed the stem pieces belong to them.

Sponge fossils also turn up occasionally. Their wrinkled surfaces are reminiscent of horn corals, but they lack the internal structures, known as septa, evident in the coral fossils.

This is one of my favorite mountain localities, as much for the ease of collecting and beauty of the Thorn Creek gorge as for the handsome and interesting fossils that are found here.

My own fossils collected from this locality are:

Brachiopods: *Cupularostrum (Camarotoechia) litchfieldense*; *Cyrtina dalmani*; *Kozlowskiellina (Delthyris) perlamellosa*; *Leptaena "rhomboidalis"*; *Meristella* (two species); *Nucleospira elegans*; *Rhipidomella* sp.; *Rhynchotreta* sp.; *Uncinulus* sp.

Bryozoans: *Fenestella* sp.; several branching species

Calcareous Algae: encrusting species

Corals: *Enterolasma (Streptolasma)* sp.; *Favosites helderbergiae*

Cystoids: stems, segments, arms of calices (brachioles)

Gastropods: *Platyceras* sp.

Poriferans (sponges): unidentified species

Stromatoporoids: unidentified species

Trilobite: unidentified tail, or pygidium

Locality 13 near Fulks Run, Rockingham County, Virginia

New Creek formation, early Devonian period

Locality 13 is a road cut and natural exposure on the east side of Route 259, 8.5 miles south of the West Virginia State line and 0.7 of a mile north of the town of Fulks Run, Virginia. There is ample space for parking on the western shoulder of Route 259 at its intersection with Route 917, directly opposite the exposure. Traffic is light to moderately heavy.

The New Creek formation (formerly known as the Coeymans formation) is one of the limestone formations included in the Helderberg group. Its fossils consist almost entirely of crinoid fragments, but the spectacle of huge boulders literally made of crinoid segments is worth seeing. Preservation is fairly poor, but recognizable segments and stem sections may be collected whole from the reddish soil beneath some of the boulders. Interesting specimens may also be obtained by cutting and polishing chunks of this rock. The white fossil content stands out against the gray matrix and makes an interesting pudding stone.

This is a good place to climb a boulder of crinoidal limestone, look at the beautiful mountain scenery, and wonder at the changes this land has seen.

The fossils I have collected are:

Brachiopods: *Desquamatia (Atrypa)* sp.
Corals: *Favosites* sp.
Crinoids: stems and segments

Locality 14 on Bullpasture Mountain, Highland County, Virginia

Corriganville formation, early Devonian period

Locality 14 is a roadside quarry and hillside on the north side of U.S. Route 250 on the eastern flank of Bullpasture Mountain. The exposures are located 4 miles west of the bridge across the Cowpasture River and 5.8 miles west of the village of Head Waters, Virginia. The best collecting is around the many large boulders that litter the hillside on the eastern fringes of the quarry. Traffic on Route 250 is light to moderate, and the shoulder is ample for parking.

The Corriganville formation, which is part of the Helderberg group, was formerly called the New Scotland formation in Virginia. Preservation of fossils is not particuarly good at this site, but the surfaces of the large limestone boulders are crowded with brachiopods, most notably, large specimens of *Macropleura macropleura* (up to three inches in width). These rocks, which are most accessible on the hillside just east of the high-walled quarry, are extremely hard and better suited for looking at or photographing than for hammering on. Fortunately, many silicified specimens have been etched free and may be gathered from the reddish sandy soil or the tops of the boulders themselves.

Most of the loose fossils to be found are fragments of brachiopods, although complete horn corals (*Enterolasma*) are fairly common, particularly in the higher parts of the exposure, near the trees. Occasional gastropods and crinoid segments also turn up. Fossils also occur in the rocks from the Keyser and New Creek formations that are exposed in the quarry proper, although collecting there is more difficult. Crinoid stems and segments are so common that some boulders are largely composed of them.

Unfortunately, the roadside areas have been planted with the ground cover crown vetch, which overwhelms much of the collecting site by midsummer with a deep carpet of greenery. This growth makes the footing treacherous and obscures most of the fossils, so plan your visits before the period of maximum growth.

This is an inspiring place to collect fossils, partly because of the size and abundance of specimens but also because the view from the mountainside as you look east toward Shenandoah Mountain is absolutely magnificent. My collection includes the following fossils from this locality:

Locality 14. Brachiopod *Macropleura macropleura*. This specimen was found on the surface of a large boulder of very hard limestone, so I collected it with a camera instead of hammer and chisel. The shell is over three inches wide.

Brachiopods: *Cyrtina varia; Eatonia medialis; Howellella (Spirifer) cycloptera; Kozlowskiellina (Delthyris) perlamellosa; Leptaena "rhomboidalis"; Macropleura (Eospirifer) macropleura; Meristella* sp.; *Platyorthis planoconvexa; Trematospira multistriata*
Corals: *Enterolasma (Streptolasma) stricta*
Crinoids: stems and segments
Gastropods: *Platyceras* sp.

Locality 15   vicinity of Smoke Hole, Pendleton County, West Virginia

Corriganville and Oriskany formations, early Devonian period

Locality 15 is 1 mile north of the town of Upper Tract and 15 miles south of Petersburg, where U.S. Route 220 crosses the South Branch of the Potomac River. Turn just south of the bridge onto the road to the Smoke Hole and Big Bend recreation areas. This road parallels the river for nearly 10 miles of a spectacular gorge, ending at the Big Bend campground. Most of the rocks exposed along the way are fossiliferous. The boulders, cliffs, and even the road gravel contain many fossils over an extensive area. The river is popular with trout fishermen and white-water canoeists, so there are plenty of pullovers along the road. Specific points of access are discussed below.

This is one of the most beautiful fossil-hunting areas I know, and the potential for exceptional finds is almost unlimited. Fossiliferous boulders are all along the scenic river and the road that follows it. But collecting is difficult, because most of the rocks are very tough to extract fossils from.

Several fossil-bearing formations are exposed in the area, including limestones of the Helderberg group and the Oriskany sandstone (known as the Ridgeley sandstone in Virginia). The Oriskany is the most conspicuous formation, comprising nearly all of the awe-inspiring outcrops that decorate the mountains and riverbanks. Its fossils are abundant and easy to spot in the countless boulders. Preservation is generally poor, but some layers with exceptionally high calcium carbonate content yield well-preserved specimens.

Specifically, a very light gray (almost white) sandstone with tiny sparkling calcite crystals visible in fresh samples is frequently packed with excellent brachiopods (*Rensselaeria, Costispirifer*) as well as parts of the giant trilobite *Trimerus*. (I've found tails indicating a total length of the whole animals in excess of eight inches!) This rock is not easy to tell from the coarser-grained layers of the Oriskany: look for the distinctive oval outlines of *Rensselaeria* fossils on the surface. The rock seems to be most common between Route 220 and the intersection and settlement 5.3 miles to the north, especially a mile or so south of the settlement.

Oriskany fossils also occur in association with chert in weathered road cuts along the same stretch of road. Reddish, sandy road banks that are covered with small pebbles and chunks of black chert are often productive. Many of these pebbles turn out to be fossils, especially gastropods (*Platyceras*), brachiopods (*Acrospirifer, Costispirifer*), tentaculitids, and crinoids. By the way, *Costispirifer arenosus* is an excellent guide fossil to the Oriskany sandstone.

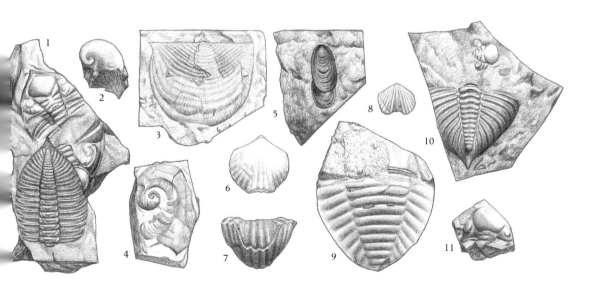

Locality 15a (Corriganville Formation). *1,* trilobites *Dalmanites pleuroptyx* (two heads and a thorax with tail attached); *2,* gastropod *Platyceras* sp.; *3,* brachiopod *Rhytistrophia beckii;* *4,* gastropod *Platyceras* sp.; *5,* brachiopod *Lingula* sp.; *6, 7,* brachiopods *Eatonia medialis;* *8,* brachiopod *Coelospira concava; 9,* trilobite *Trimerus vanuxemi* (tail); *10,* trilobites *Dalmanites pleuroptyx* (tail and partial head); *11,* trilobite *Phacops logani* (head). (Fossils shown .6x actual size.)

My best collecting in the Helderberg formations exposed in this area has been in the tough, very dark gray limestones of the Corriganville formation. The Corriganville is best exposed along the road between the settlement and Big Bend campground. Some chunks are very fossiliferous, full of brachiopods (*Eatonia, Leptaena*). These are hard to extract, as the rock is resistant to hammer blows, and when it does break open the fossils are usually damaged.

My best brachiopod fossils from this type of rock actually came from locally derived road gravel near Big Bend. Sometimes, whole specimens may be found that are better looking than those retrieved from fresh rocks, despite having been run over repeatedly. Many specimens are covered with intricate patterns, consisting of minute concentric circles and wavy lines. These are the fossils of encrusting calcareous algae (which also occur on some of the better preserved Oriskany fossils).

The road to Big Bend crosses a small stream 2.1 miles north of the store and settlement. An especially thick-bedded, chert-free form of this dark gray Corriganville limestone crops out along the stream, about two hundred yards west of the road. This rock is so hard it rings when you hit it with a hammer and sends dangerous splinters flying, but it contains many fossils of the large trilobite *Dalmanites pleuroptyx,* especially tails. I've found isolated tails as much as two inches long, so a five-inch complete animal is probably hiding somewhere. It is safer to crack these rocks with a series of small blows than with one mighty one, so that flying slivers will have less force behind them.

Larger still are the fossils of the trilobite *Trimerus,* the same kind that occurs in the Oriskany sandstone nearby. I've found a couple of large tails in this same gray limestone—in road gravel, in fact. *Trimerus* is cylindrical in shape and much heavier than the flattened *Dalmanites.*

I could spend a lifetime collecting here—and it would take a lifetime just to figure out the rocks. So much material is present, especially since the catastrophic flood of November 1985 and the subsequent road repairs, that it is overwhelming to a fossil collector used to tidy little road cuts. The rocks contain giants, but you have to work for them.

These are the fossils I have collected from the Corriganville formation:

Brachiopods: *Coelospira (Anoplotheca) concava; Cyrtina varia; Eatonia medialis; Howellella (Spirifer) cycloptera; Leptaena* "*rhomboidalis*"; *Lingula* sp.; *Meristella* sp.; *Platyorthis planoconvexa; Rhytistrophia beckii* (similar to *Leptaena* but more flattened)
Bryozoans: encrusting and massive species
Calcareous algae: encrusting species
Corals: *Favosites helderbergiae*
Crinoids: stems and segments
Gastropods: *Platyceras* sp.
Trilobites: *Dalmanites pleuroptyx; Phacops logani; Trimerus vanuxemi*

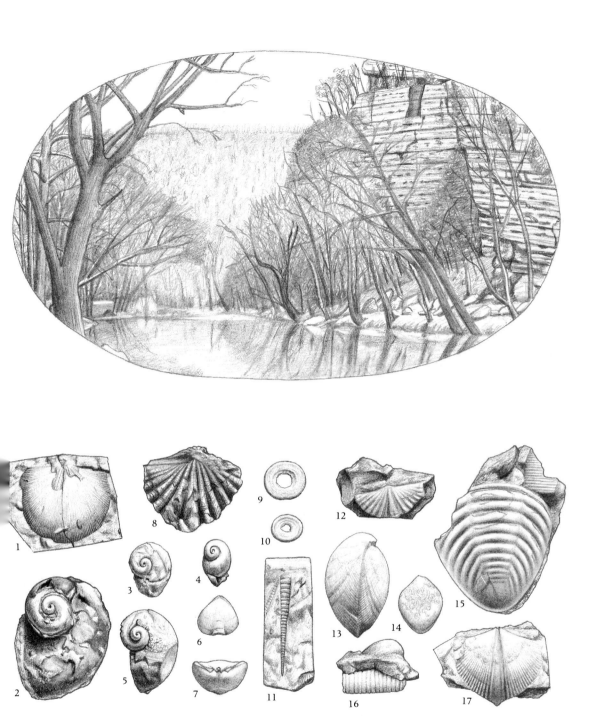

Locality 15b (Oriskany Formation). *1*, brachiopod *Schellwienella* (?) sp.; *2–5*, gastropods *Platyceras* sp.; *6, 7*, brachiopods *Costellirostra peculiaris*; *8*, brachiopod *Acrospirifer murchisoni*; *9, 10*, crinoid segments; *11*, tentaculitid *Tentaculites* sp.; *12*, brachiopod *Acrospirifer murchisoni*; *13, 14*, brachiopods *Rensselaeria* sp. (note calcareous algae on *14*); *15*, trilobite *Trimerus vanuxemi* (tail); *16*, crinoid stem section; *17*, brachiopod *Costispirifer arenosus*. (Fossils shown .6x actual size.)

These are the fossils I collected from the Oriskany formation:

Brachiopods: *Acrospirifer murchisoni; Costellirostra peculiaris; Costispirifer (Spirifer) arenosus; Cupularostrum (Camarotoechia)* sp.; *Cyrtina varia; Meristella lata; Rennselaeria* sp.; *Schellwienella* (?) sp.
Calcareous algae: encrusting species
Crinoids: stems, segments, partial crown (calyx)
Gastropods: *Platyceras* sp.
Trilobites: *Dalmanites* sp.; *Trimerus vanuxemi*

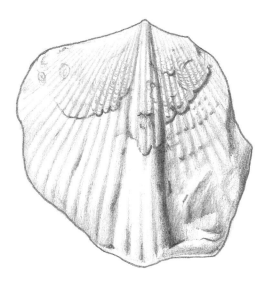

Locality 15c (Oriskany Formation). Brachiopods *Costispirifer arenosus*. (Note the encrusting calcareous algae on the upper portion of the lower brachiopod.) This species is an excellent guide fossil to the Oriskany-Ridgeley sandstones and is extremely common at this locality. (Specimens shown 1.2x actual size.)

Locality 16 — near Monterey, Highland County, Virginia

Licking Creek formation, early Devonian period

Locality 16 fossils are found in the debris on a road cut and hillside 11.1 miles south of the junction of U.S. routes 220 and 250 in the town of Monterey, Virginia, on the west side of Route 220, immediately north of its intersection with Route 606. Route 606 crosses the Jackson River just east of its junction with Route 220. Parking is best on the wide shoulder directly across Route 220 from the Route 606 bridge. Collecting begins about 100 yards north (uphill) from this parking area.

This is another of my favorite mountain collecting sites, because the fossils are fairly large, undistorted in shape, varied in type, and most of all, they can be collected with ease. This last virtue is due to the fact that the specimens are silicified, have been weathered free from the limestone that contained them, and are scattered loose on the surface of the soil. In other words, easy pickings. But do your collecting in the late fall to early spring, because the fossils are obscured by summer's heavy vegetation, which includes nettles and other thorny plants.

Among the most distinctive fossils here are the dome-shaped colonial corals *Favosites conicus*. The flattened bases of these fossils allow them to sit nicely in a display case. These corals are sometimes encrusted with the coral fossil *Aulopora*.

Many beautiful brachiopods are abundant, especially complete specimens of *Rensselaeria, Meristella, Costellirostra,* and *Eatonia*. Two of the *Eatonia* specimens I found were small geodes with sparkling white quartz crystals lining their hollow interiors. Another complete, hollow brachiopod specimen (*Costellirostra*) came apart with a little coaxing into its two valves, or shells. This allowed us to see the inner structure, including the brachidium, or the solid support on one of the valves for the soft filter-feeding organ (lophophore), which is attached in the living organism.

Another unusual fossil from this fine locality is the cuplike or vaselike crown of the crinoid *Edriocrinus*. It was difficult to visualize these relatively featureless, thimble-shaped specimens as parts of crinoids until I found one with its pentagonal top intact, to which five branching arms had been attached in life.

Fossils are strewn over the hillside from top to bottom, with a concentration of the larger corals in the gully at its base. The limestone from which they derive crops out here and there, and some chunks of weathered rocks consisting primarily of fossil shells cemented together may be found. About a quarter mile to the north, at the top of the hill, more fossils occur, including abundant stromatoporoid specimens.

The only problem with collecting at this locality is that it makes it hard to go back to pounding on the stubborn rocks that grip most of the fossils in other localities in these mountains.

My own collection from locality 16 includes:

Brachiopods: *Costellirostra (Eatonia) singularis; Cyrtina varia; Eatonia medialis; Howellella (Spirifer) cycloptera; Leptaena "rhomboidalis"; Meristella lata; Platyorthis planoconvexa; Rensselaeria subglobosa; Rhipidomella assimilis; Strophonella* sp.

Bryozoans: encrusting and massive species

Calcareous algae: encrusting species

Corals: *Aulopora* sp. (encrusting species); *Favosites conicus* (hemispheric colonies); *Enterolasma (Streptolasma) stricta* (horn coral)

Crinoids: *Edriocrinus pocilliformis* (calices); incomplete calyx of unidentified species

Cystoids (?): stems and segments

Stromatoporoids: unidentified species

Trilobites: *Dalmanites* sp.

Locality 16. *1,* stromatoporoid; *2,* bryozoan colony; *3,* coral *Enterolasma stricta; 4,* corals *Aulopora* sp. growing on the underside of a colony of the coral *Favosites conicus; 5,* coral *Favosites conicus* (side view); *6,* calcareous algae; *7,* brachiopod *Costellirostra singularis; 8, 9,* brachiopods *Rensselaeria subglobosa; 10,* brachiopod *Meristella lata; 11, 12,* brachiopods *Rhipidomella assimilis* (*12* shows the interior of a valve); *13,* brachiopod *Eatonia medialis; 14,* brachiopod *Howellella cycloptera; 15, 16,* cystoid (?) stem and segment; *17,* crinoid partial calyx; *18, 19,* crinoids *Edriocrinus pocilliformis* (calices); *20a–20b,* brachiopod *Cyrtina varia* (two views of a single valve). (Fossils shown .6x actual size.)

Locality 17 near the Lost River, Hardy County, West Virginia

Needmore formation, early to middle Devonian period

Locality 17 is a roadside quarry on the north side of Route 55-259, 4.2 miles west of the town of Wardensville, West Virginia, and about 0.3 of a mile west of the bridge over the Lost River. There is a wide shoulder at the quarry with plenty of parking space, but be careful entering or leaving the highway; traffic can be moderately heavy, the exposure is at the crest of a steep hill, and visibility is poor.

A quick stop at this locality could lead one to believe it has little potential for fossil hunting. Many of the shales exposed are very fissile (prone to splitting into flakes and sheets), and most of the fossils on their surfaces are small and difficult to collect intact. But I know of few Paleozoic localities that can boast of a greater variety of fossils, including seven species of trilobites.

One of the reasons for the diversity of forms here is the diversity of rock types, each representing different underwater conditions during deposition of sediments. There are very dark gray shales with small brachiopods and straight-shelled nautiloids, either of which may be pyritized (meaning they are made of pyrite, or fool's gold). There are light gray to green shales with trilobites (*Phacops cristata, Dalmanites, Odontopleura*), more brachiopods (*Ambocoelia, Desquamatia*), large pelecypods (*Praecardium*), and crinoids. And there are yellow shales and orange decalcified limestones with abundant trilobites (*Phacops rana, Odontocephalus*), brachiopods (*Desquamatia, Coelospira*), and horn corals.

The layout of the quarry is subject to change, as it is still used by highway workers, but it has remained pretty much as follows for several years. There are two large bays separated by a central heap of debris, which comes almost to the road. The western bay is faced by a nearly vertical exposure of gray and green shale. If you look closely at the surface of these rocks, you will see many small fossils scattered across them. Keep an eye out for trilobites, especially *Phacops cristata* (distinguished from its more widespread relative, *Phacops rana,* by its longi-

Locality 17a *(opposite).* 1, 2, trilobites *Odontocephalus aegeria;* 3, 4, trilobites *Odontopleura callicera* (3 is a partial head, 4 is a partial thorax and tail); 5, 6, trilobites *Phacops cristata* (5 is a head, 6 is a thorax and tail; note spines on middle lobe); 7, horn coral; 8, brachiopod *Rhipidomella vanuxemi;* 9, trilobite *Phacops rana* (note absence of spines on middle lobe); 10, pelecypod *Praecardium multiradiatum;* 11, gastropod *Loxonema* sp.; 12, crinoid stem section; 13, 14, nautiloids *Michelinoceras subulatum;* 15, trilobite *Dalmanites* sp. (Fossils shown .6x actual size.)

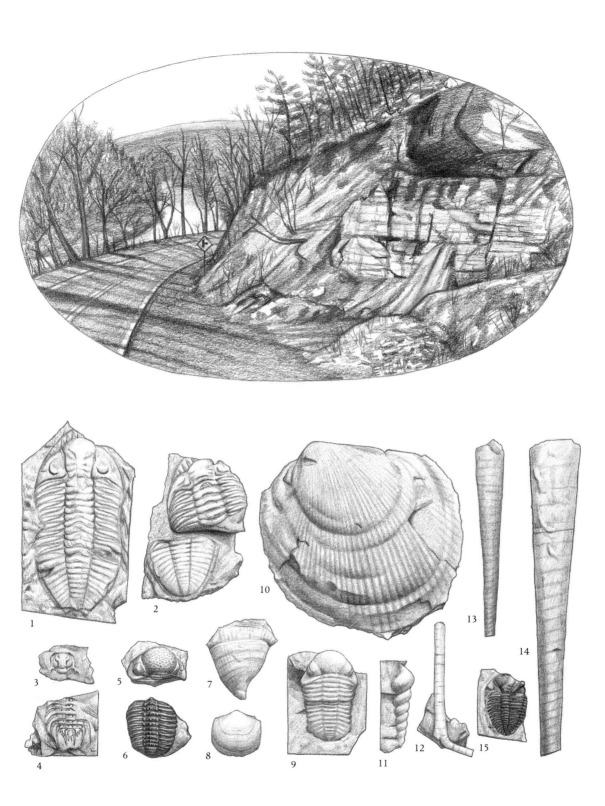

tudinal row of spines on the center lobe). Many of the trilobites in these rocks have their shells, or exoskeletons, intact—usually in black calcite. Careful cleaning with a brass-bristled brush can yield spectacular results. Trilobites with this kind of preservation are easiest to prepare when collected from exposed surfaces, since the right amount of weathering makes them easier to retrieve and clean up.

The straight-shelled nautiloids in the dark gray shale attain at least seven or eight inches in length. Some are partially pyritized. They almost always break during collecting or are broken when you find them. Collect all the pieces, and Elmer's Glue-all will work magic for you. The same goes for crinoid stems.

Beginning at the center of the quarry, just above the pile of debris and extending into the eastern bay, is a nearly vertical layer of yellowish shale, which grades into an orange, chalky rock as you go deeper. These rocks contain many trilobites; most of them are complete, but they are not easy to collect. The rock is very soft, especially the orange layer, and the fossils are often oriented against the grain. With patience, however, and with some at-home preparation, you will almost certainly acquire some nice specimens.

Earlier, I referred to the orange rock as decalcified limestone. Limestone is largely calcium carbonate, which dissolves in acidic rainwater. The limestone in this layer has a very high content of mud and silt, so when the calcium carbonate was dissolved away, a soft, clayey rock was left behind. The shells of the trilobites were dissolved away as well, so these specimens are internal and external molds. Recent excavations have exposed some fresher samples of this limestone, which have not been decalcified. This rock is greenish and very hard, and its fossils are very difficult to extract.

I had not fully appreciated the yellow and orange layers until recently, since early inspections of this part of the quarry had never turned up much more than headless, poorly preserved *Phacops rana* fossils in the surface shales. It wasn't until I dug into the cliff that I hit pay dirt. After freeing one forty-pound chunk, I gave it to my brother Phil and said, "I'll give you a 99 percent guarantee there's a complete trilobite in this rock." I was wrong—there were *five*! One difficulty with collecting from this layer is the extreme softness of the rock; it can actually be broken by hand and gouged by fingernails. Do your collecting and packing up very gently, and immerse the specimens in diluted Elmer's Glue-all when you get home.

A similar layer with the same kinds of fossils is exposed at locality 20. But the limestone layers there have not been decalcified, so the trilobites are nearly impossible to extract except from the surrounding shales. At both localities, the fossiliferous layers are overlain by the twisted, fissile, unfossiliferous shales of the Marcellus formation. These rocks are exposed at the eastern edge of this locality.

While you're here, notice the white Oriskany sandstone exposed at the western edge of the quarry. It contains a few fossils, and other exposures of the same formation in the area contain more. Also, listen for the Lost River, across the road and down the hill from the quarry. The Lost River is so named because a stretch of

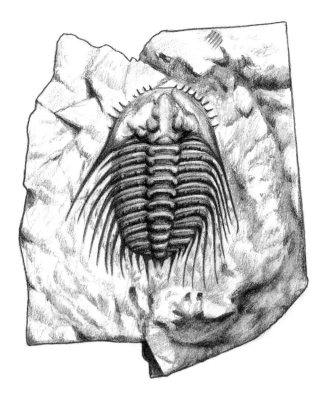

Locality 17b. Trilobite *Odontopleura callicera*. The fossil itself is one and one-eighth inches long.

it disappears for most of the year into an underground course, starting just above the bridge that crosses it a quarter mile east of the locality. Two miles downstream it reemerges and picks up a new name, the Cacapon.

Fossils I have collected at this locality are:

Brachiopods: *Ambocoelia nana; Coelospira (Anoplotheca) acutiplicata; Desquamatia (Atrypa) reticularis; Eodevonaria arcuata; Leptaena "rhomboidalis"; Rhipidomella vanuxemi; Schuchertella pandora*
Bryozoans: *Fenestella* sp.
Cephalopods: *Agoniatites vanuxemi* (coiled ammonoid); *Michelinoceras subulatum* (straight-shelled nautiloid)
Corals: *Pleurodictyum* sp. (disc-shaped colonies); *Trachypora* (?) sp. (branching colonies); three species of solitary horn corals
Crinoids: stems and segments
Gastropods: *Loxonema* sp.; *Platyceras* sp.; *Platyostoma* sp.
Pelecypods: *Modiomorpha* sp.; *Praecardium (Panenka) multiradiatum*
Pteropods: needle-shaped, gastropodlike mollusks
Trilobites: *Basidechenella (Proetus) rowi; Coronura aspectans; Dalmanites* sp.; *Odontocephalus aegeria; Odontopleura callicera; Phacops cristata; Phacops rana*

Locality 18   near Capon Lake, Hampshire County, West Virginia

Needmore formation, early to middle Devonian period

Locality 18 is a small roadside quarry on the west side of Route 259, 0.2 of a mile south of the village of Capon Lake and 5.2 miles north of the town of Wardensville, West Virginia. The shoulder broadens at the quarry, providing ample parking.

The Needmore foundation was formerly called the Onondaga by local geologists. Many of the older names for formations in the area were derived from the similarity of their fossils with those of formations of approximately the same age in New York State, where these assemblages of fossils were first described by scientists. However, formations are distinguished on the basis of more than fossil content, so most of these names have been changed to reflect the regional differences in rock types, geologic history, and the lack of geographic continuity with the New York rocks. These changes should be borne in mind when using older source materials.

The fossils at this locality are similar in types and preservation to those found in the shales at locality 17. Most of the rocks exposed at the southern end of the quarry are olive green shales, but the northern three-quarters of this locality contains shales ranging in color from gray to yellow, cream, pink, and orange. All colors of rock contain fossils.

Very small ostracode fossils are abundant here, although it takes some looking to notice. A close inspection of some of the lighter colored rocks reveals a host of specimens, most of them a millimeter or less in length—although I did find one that is four and a half millimeters long. Most specimens are small, with the exception of some very large examples of the pelecypod *Praecardium multiradiatum*. The variety of specimens is excellent, but it takes a gentle touch to retrieve many of them, as the rocks are soft and the beds are fairly thick and reluctant to break into even layers when fresh.

Recent highway construction has made a great deal of fresh rock available but has also made collecting more dangerous, since the hillside on the northern part of the outcrop is a very unstable heap of boulders. Be alert for unexpected rockslides, and climb with care.

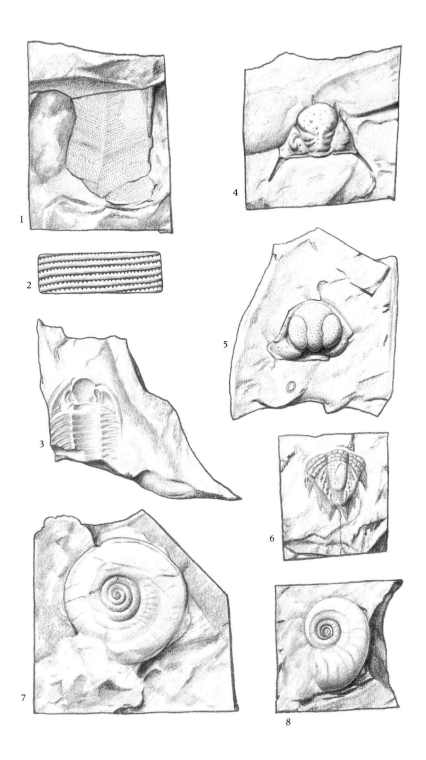

Locality 18. *1, 2,* conularid *Conularia* sp. (*2* is shown 8.5x actual size); *3,* trilobite *Basidechenella rowi; 4,* trilobite *Dalmanites* sp. (head); *5, 6,* trilobites, unidentified species (*5* is a partial head, *6* is a tail); *7, 8,* ammonoids *Agoniatites vanuxemi*. (Specimens other than *2* shown 1.4x actual size.)

Fossils collected here include:

Brachiopods: *Ambocoelia nana; Coelospira (Anoplotheca) acutiplicata; Desquamatia (Atrypa) reticularis; Leptaena "rhomboidalis"*
Cephalopods: *Agoniatites vanuxemi* (coiled ammonoid); *Michelinoceras subulatum* (straight-shelled nautiloid)
Conularids: *Conularia* sp.
Corals: solitary horn corals
Crinoids: stems and segments
Gastropods: *Loxonema* sp.
Ostracodes: *Bollia* sp.; *Kloedenia*(?) sp. (large)
Pelecypods: *Lyriopecten* sp.; *Nuculoidea* sp.; *Praecardium (Panenka) multiradiatum*
Trilobites: *Basidechenella (Proetus) rowi; Dalmanites* sp.; *Odontocephalus aegeria; Odontopleura callicera; Phacops cristata; Phacops rana;* unidentified species (similar to *Lichas*)

Locality 19 — east of Franklin, Pendleton County, West Virginia

Needmore formation, early to middle Devonian period

Locality 19 is a fairly large road cut–roadside quarry on the south side of Route 33, 8.4 miles east of the junction of U.S. routes 33 and 220 in Franklin. There is a mailbox in the shoulder near the middle of the quarry. The shoulder is wide enough to permit parking. There is also a flat area between the road and the outcrop, but it is strewn with boulders that may be hidden by vegetation. Traffic is moderate, but the steep grade of the highway and speeding trucks can make coming and going a bit tricky.

Most of the rocks in this quarry are extremely fissile (prone to splitting into thin sheets), dark gray to black shales. These rocks break apart very easily and contain pieces of pyrite as well as some small pyritized nautiloid fossils. Some layers of more substantial green or brown shale occur, which are frequently packed with flattened trilobite fossils (*Phacops cristata* and *Phacops rana*). But the best specimens come from the irregularly shaped, dark gray or green limestone nodules that occur within the shale layers. These contain numerous specimens of *Phacops cristata* with their exoskeletons intact, usually in black calcite, which shines after it is cleaned with a brass-bristled brush.

I haven't been very successful in locating the sources of the better rocks in the cliff itself; the lower layers consist primarily of the fissile shales, and climbing on the sheer, fractured surface of the cliff is hazardous. But enough rock is scattered at its base to keep one busy. It is likely that the best layers continue in the roadcut west of the quarry, where they may be more accessible. I have searched there and found a few specimens but never enough to keep me away from the rockfall on the eastern edge of the quarry, where nearly all of my best fossils have turned up.

Some of the pyritized nautiloids have their delicate internal structures preserved. These details are also visible on a specimen that was inside one of the chunks of pyrite, which I cut in half with my rock saw and polished. But if you do saw pyrite, be prepared for a mess; pyrite dust plus water equals a black inky muck, which is difficult to wash off skin and clothing.

I know there are some real museum pieces, especially of the trilobite *Phacops cristata*, to be found at this locality. I haven't found them yet. Perhaps you will. Here are the fossils I have collected:

Brachiopods: *Coelospira (Anoplotheca) acutiplicata; Desquamatia (Atrypa) reticularis*

Cephalopods: *Michelinoceras* sp.; *Spyroceras* sp. (straight-shelled nautiloid with wavy profile)

Corals: *Heterophrentis* sp. (horn coral); *Trachypora*(?) sp. (branching colonies)

Crinoids: segments

Gastropods: *Loxonema* sp.

Pelecypods: *Praecardium (Panenka) multiradiatum*

Trilobites: *Phacops cristata; Phacops rana*

Locality 20 along the Cacapon River, near Yellow Spring, Hampshire County, West Virginia

Needmore formation, early to middle Devonian period

Locality 20 consists of a large road cut and natural exposures on the west bank of the Cacapon River along the Capon River Road (Route 14), 2 miles north of its intersection with Route 259 in the village of Yellow Spring. Fossils occur in the boulders along the river as well as in the road cut above. (Capon River Road continues north for 5 or 6 miles and intersects with Route 50 in the town of Capon Bridge.)

The shoulders are narrow and the hill steep, so choose your parking space carefully. There is a fairly wide grassy shoulder about a hundred yards north of the best collecting, just south of a Speed Zone Ahead sign and before the rocks become very thin bedded, distorted, and unfossiliferous shales from the Marcellus formation.

All of the fossiliferous rocks at this locality are medium to dark gray. The layers above the road are tilted toward the river, which makes climbing difficult and treacherous. Easier collecting may be done along the shore of the river in the many loose boulders, although the climb down from the road is no piece of cake, either. The best and safest hunting is in the easily excavated rocks at road level.

The southern outcrops along the road consist of gray, fissile shale. Weathered surfaces of this rock are broken into small pieces, and so are the fossils; but the fresher material contains many superb specimens, including complete specimens of the trilobite *Phacops rana* and pieces of the large, ostentatiously ornate trilobite *Coronura aspectans.* You will have to remove and dissect fresh slabs of this shale in order to collect these fossils in good condition.

Excellent fossils also occur in a slab of hard, blocky limestone (eight to ten inches thick) that covers a large area of the road cut, just south of the wooded slope. The layer itself is difficult to climb and extremely hard to extract fossils from, but the shales just above and below it contain many fine specimens, especially of the trilobites *Odontocephalus* and *Phacops,* often with the exoskeleton preserved in black calcite (crystalline calcium carbonate). Whole specimens of the brachiopods *Coelospira, Desquamatia, Rhipidomella,* and others also occur. These are not internal molds; they include the shells and surface details preserved in silvery calcite.

Be very careful to avoid starting avalanches here, as nearly everything winds up on the road below. Traffic is usually light, but cars and trucks have little room to avoid obstacles.

The fossils I have collected at this locality are:

Brachiopods: *Ambocoelia nana; Coelospira (Anoplotheca) acutiplicata; Devonochonetes hemisphericus; Desquamatia (Atrypa) reticularis; Leptaena "rhomboidalis"; Orbiculoidea lodiensis; Rhipidomella vanuxemi; Schuchertella pandora*

Bryozoans: *Fenestella* sp.

Cephalopods: *Agoniatites vanuxemi*

Corals: *Heterophrentis* sp.; *Pleurodictyum* sp.; *Trachypora*(?) sp.; *Zaphrentis* sp.; unidentified small, squat horn coral

Crinoids: segments and stems

Gastropods: *Platyceras* sp.; *Platyostoma* sp.

Hyolithids: *Hyolithes* sp.

Pelecypods: *Praecardium (Panenka) multiradiatum*

Plants: coal fragment in shale

Trace fossils: abundant trails on rock surfaces

Trilobites: *Coronura aspectans; Odontocephalus aegeria; Phacops rana*

Locality 20 *(opposite)*. *1,* crinoid stems; *2,* coral *Trachypora* (?) sp.; *3,* horn coral; *4,* horn coral *Heterophrentis* sp.; *5,* horn coral *Zaphrentis* sp.; *6,* hyolithid *Hyolithes* sp.; *7,* crinoid stem; *8–10,* trilobites *Phacops rana; 11,* brachiopod *Schuchertella pandora; 12, 13,* trilobites *Odontocephalus aegeria (12* is the head and thorax, *13* is the tail); *14,* brachiopod *Coelospira acutiplicata; 15,* brachiopod *Desquamatia reticularis; 16,* brachiopod *Rhipidomella vanuxemi; 17, 18,* trilobites *Coronura aspectans (17* is the head and thorax, *18* is the tail). (Specimens shown .6x actual size.)

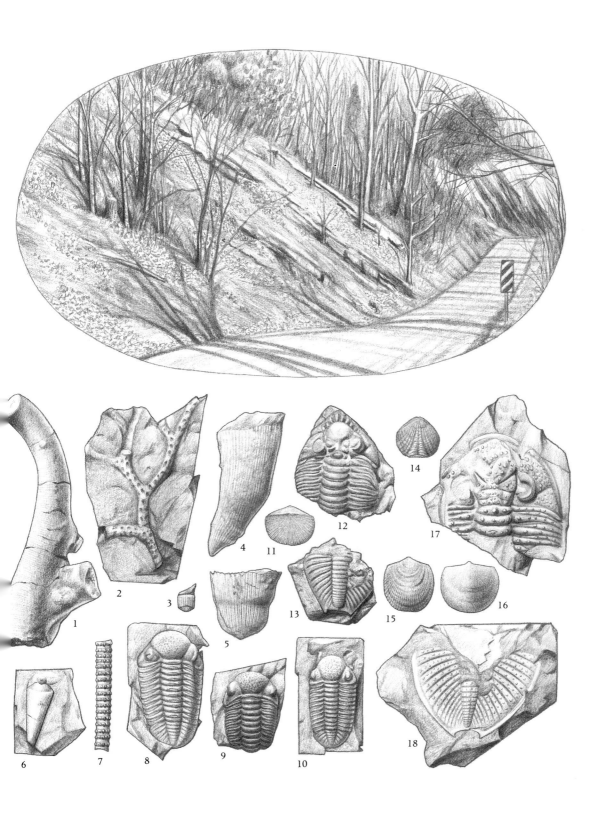

Locality 20, Cacapon River, West Virginia

Locality 21 — near Fort Frederick State Park, Washington County, Maryland

Marcellus formation, middle Devonian period

Locality 21 is a low roadside quarry on the north side of Route 56, 1.8 miles east of the entrance to Fort Frederick State Park, 2.1 miles west of Big Spring and 3.2 miles east of Interstate Route 70 exit 12 (to Big Pool and Indian Springs). *This locality is on private land and permission to collect must be obtained in advance.* The owner's name is on a No Trespassing sign at the locality, and he may be contacted by calling directory assistance. There is a wide parking area between the exposure and the road.

This is an unusual collecting locality. Most of the rocks here are very fissile, gray and green shales, which split readily into thin sheets. These layers contain very few fossils, but those few fossils are extremely interesting, being magnificently preserved external molds of crinoids. About one hundred yards north of the parking area, in the northeastern corner of the quarry, are layers of a dark gray limestone that weathers to a light gray or tan. These rocks are hard to find and are not abundant but are far more productive than the shales, containing numerous examples of the crinoid *Arthroacantha punctobrachiata*.

Some of the specimens at this locality have the crinoid parts still intact as calcite, but these are not nearly as satisfactory as those from which the calcite has been dissolved away. The calcite material does not contrast well with the fresh matrix and cannot be fully exposed by mechanical means. By comparison, the external molds stand out, especially in the lighter colored rocks, thanks to iron oxide residues, and preserve the structure of the delicately branching animals in minute detail.

I know of no other locality in which complete crinoids so regularly occur. Still, it takes some looking to find them. Isolated stems and arms of the crowns are fairly common, however. Fossils from the overlying Mahantango formation may be found in exposures along the road to McCoy's Ferry, which enters Route 56 from the south opposite the locality.

I have collected:

Brachiopods: *Cyrtina* sp.; *Spinocyrtia granulosa*
Crinoids: *Arthroacantha punctobrachiata*; *Lasiocrinus scoparius*
Plants: fragments

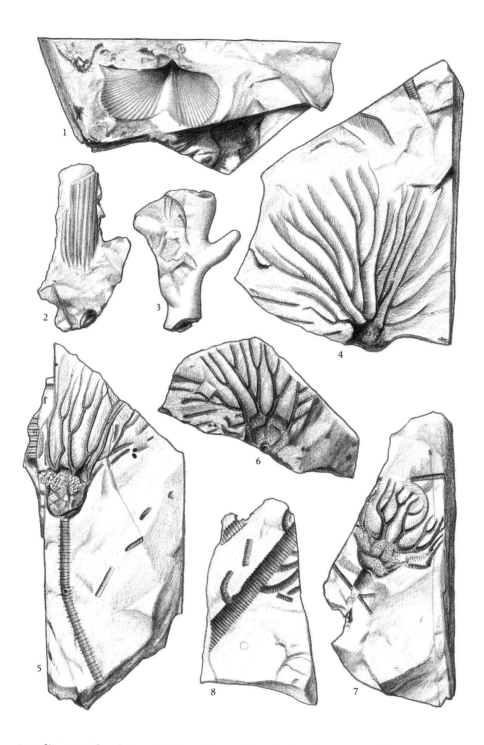

Locality 21. *1*, brachiopod *Spinocyrtia granulosa*; *2*, plant fragment; *3*, branching crinoid stem surrounded by concretion; *4–7*, crinoids *Arthroacantha punctobrachiata* (*4* and *6* are isolated crowns, or calices; *5* and *7* have parts of the stem attached); *8*, crinoid stem. (Specimens shown .7x actual size.)

Locality 22 Gore, Frederick County, Virginia

Mahantango formation, middle Devonian period

Locality 22 is a road cut on the north side of Route 50 directly opposite the town of Gore, which is south of the highway. The exposure is 500 feet west of the Hebron Baptist Church and 11 miles west of Winchester, Virginia. The shoulder at the western end of the road cut is just wide enough for parking. Route 50 is a busy, four-lane highway, so be careful not to litter the road with fallen rocks—or fossil collectors.

This is one of the very few mountain localities listed here that I know to be frequented by collectors. And for good reason; it is probably the most reliable source of trilobite fossils in the area.

Only one species has turned up for me, and that is *Phacops rana.* But has it ever turned up! I've filled a drawer with scores of specimens, including about a dozen complete ones. Most of the fossils are internal molds, and a high percentage are isolated heads and headless rollers, or the enrolled thorax and tail. These are believed to be the products of molting—the shedding of skins—rather than the remains of dead trilobites. Maybe one complete trilobite is found for every thirty or forty partial specimens, and nine of ten complete fossils are enrolled.

From my experience, the best collecting is at the western end of the road cut. The amount of debris in this area indicates that most collectors agree with me; but I did find an unusually well-preserved, nearly complete specimen lying on the surface of the cliff at about the middle of the exposure. It might have been dropped by another fossil hunter, though, as I have found no more there.

Fossils occur in dark brown to tan and light green shale, siltstone, and fine-grained sandstone. The most productive layers are most accessible at the very top of the road cut, but these rocks are also more weathered, and the fossils less well preserved. Better looking fossils turn up in fresher, dark brown rock, which can be very tough to crack open.

As wave after wave of collectors have dug into the top of the cliff, the lower part of the exposure has been covered by a blanket of debris. This has not only

Locality 22 *(opposite).* 1, brachiopod *Tropidoleptus carinatus;* 2, pelecypod internal mold; 3–5, crinoids *Arthroacantha* sp. (3 is a stem, 4 is a cast, or internal mold, of a calyx, 5 is a stem with calyx attached); 6, gastropod *Platyceras* sp.; 7, brachiopod *Spinocyrtia granulosa* (internal mold); 8–11, trilobites *Phacops rana* (9 and 10 are enrolled); 12, gastropod *Bembexia ella* (internal mold). (Fossils shown .6x actual size.)

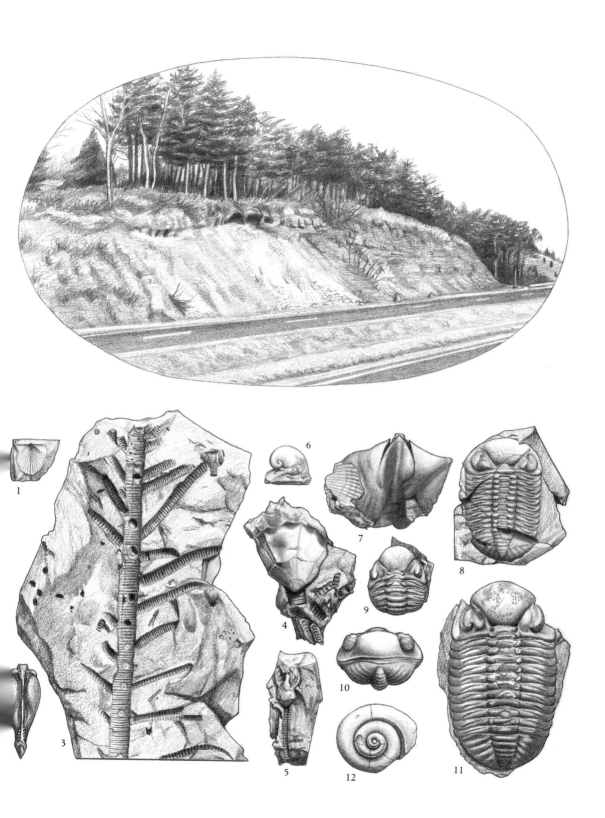

Locality 22, Gore, Virginia

covered much of the fresher material below, it has also made the climb to the top much more difficult. Choose the route for your ascent and descent very carefully, hanging onto the small trees.

Every time I approach this locality, I half expect to see it posted with No Parking signs. The debris comes very close to the road sometimes, and there is an element of danger in collecting so close to the highway. But I hope no irresponsible collecting results in its closure, because this is an unusually rich source of the elusive trilobite. My own finds are listed below. Many more fossils than these occur—I confess to being preoccupied with the trilobites.

Brachiopods: *Protoleptostrophia (Stropheodonta) perplana*; *Spinocyrtia granulosa*; *Tropidoleptus carinatus*
Cephalopods: straight-shelled nautiloids
Crinoids: segments, stems, and crowns; *Arthroacantha* sp.
Gastropods: *Bembexia (Pleurotomaria) ella*; *Loxonema delphicola*; *Platyceras* sp.
Pelecypods: internal molds of several species; *Modiomorpha* sp.
Trilobites: *Phacops rana*

Locality 23 east of Wardensville, Hardy County, West Virginia

Mahantango formation, middle Devonian period

Locality 23 consists of extensive road cuts on the north side of Route 55, especially 1.8 miles east of the town of Wardensville, West Virginia, and 1.8 miles west of the Virginia State line. The shoulders on both sides of the road are wide enough in several places to permit parking, especially at the recess at the east end of the best exposure near a small quarry. Traffic is light to moderate.

Several road cuts along the highway for a mile or so expose a variety of rock types belonging to the Mahantango formation. Typically, the lower beds of the Mahantango are sparsely fossiliferous black shales, which give way to coarser-grained, more fossiliferous green and gray shales and siltstones as you ascend through the layers, culminating in highly fossiliferous gray or light brown siltstones and sandstones at the top of the formation. This sequence is visible here from west to east, in ascending order.

The best collecting at this locality is in the gray and green shales in the middle of the formation and in the dense, extremely hard limestone nodules that occur in these layers. These are exposed in the road cut just to the east of the local road that enters the highway from the north. Fossils also occur in abundance in the upper sandstone layers, which are exposed in the next outcrop to the east (uphill) but are generally less well preserved.

As at other localities where fossils occur in hard nodules, be careful when splitting these rocks open. Complete brachiopods may be found within them (*Spinocyrtia, Ambocoelia*) as well as occasionally pyritized specimens of the ammonoid *Agoniatites*. These nodules (known as concretions) are most abundant around the small quarry on the eastern edge of the best road cut mentioned above. Some enormous boulders in this area also contain interesting fossils, including very large nautiloids, corals (*Pleurodictyum*), and well-preserved brachiopods (*Mucrospirifer*), but don't expect them to yield their fossils easily. These rocks are remarkably hard to impress with hammer and chisel, as I discovered to my dismay when I damaged a large example of the unusual brachiopod *Lindstromella*.

Easier collecting may be done in the gray shales exposed just to the west in the road cut. These contain many brachiopods as well as crinoid stems and a host of pelecypods (*Nuculites, Modiomorpha, Palaeoneilo*, etc.). A bit further west are some layers of calcareous shale, which are crowded with fossils. These specimens are best appreciated in samples from which the shell material has been removed by weathering. Most common are examples of the elegant brachiopod *Mucrospiri-*

fer mucronatus, but many other fossils occur, including other brachiopods, bryozoans, trilobites, and gastropods.

The fossils I have collected here are:

Brachiopods: *Ambocoelia umbonata; Devonochonetes scitulus; Elita (Elytha) fimbriata; Lindstromella aspidium; Lingula* sp.; *Mucrospirifer mucronatus; Orbiculoidea* (?) sp.; *Spinocyrtia granulosa; Tropidoleptus carinatus*
Bryozoans: *Fenestella* sp.
Cephalopods: *Agoniatites* sp. (coiled ammonoid); *Michelinoceras* sp. (straight-shelled nautiloid)
Corals: *Favosites* sp.; *Heterophrentis* sp.; *Pleurodictyum* sp.; *Trachypora* sp.
Crinoids: stems, segments, crowns; *Lasiocrinus* (?) sp.
Gastropods: *Bembexia (Pleurotomaria) ella; Bucanella* sp.; *Crenistriella* sp.; *Loxonema hamiltoniae*
Pelecypods: *Cornellites flabella; Grammysia bisulcata; Grammysiodea alveata; Modiomorpha concentrica; Nuculites oblongatus; Nuculoidea* sp.; *Orthonota* sp.; *Palaeoneilo* sp.
Tentaculitids: unidentified species
Trilobites: *Greenops boothi; Phacops rana; Trimerus (Dipleura) dekayi*

Locality 23 *(opposite).* 1, pelecypod *Cornellites flabella;* 2, coral *Pleurodictyum* sp.; 3, brachiopod *Mucrospirifer mucronatus.* (Specimens shown 1.5x actual size.)

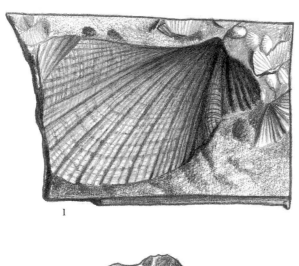

1

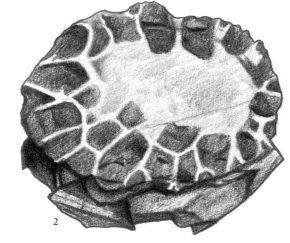

2

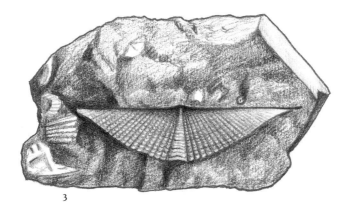

3

Locality 23, Wardensville, West Virainia

Locality 24 — near Baker, Hardy County, West Virginia

Mahantango formation, middle Devonian period

Locality 24 is a series of road cuts on the west side of Route 259, stretching from 4 to nearly 5 miles south of its intersection with Route 55-259 in the town of Baker, West Virginia. The shoulders are wide enough for parking along the road, either adjacent to the road cuts or across the highway. Traffic is light to moderate.

There are actually three localities here, but I have listed them together because of their proximity to each other. Each has a distinct assemblage of fossil types, although some species occur at all three exposures.

The northernmost road cut contains many large examples of the brachiopod *Spinocyrtia granulosa*. Some of these specimens are more than three inches wide, and many still retain white calcite shell material, although fossils consisting of casts and impressions are more easily collected. Many pelecypod specimens also occur here, including *Modiomorpha, Cypricardella,* and large *Ptychopteria* fossils. An occasional fragment of the trilobite *Greenops boothi* also turns up.

About a half mile farther south at the next large outcrop, a much more diverse fossil fauna may be collected. Both colonial and solitary corals, bryozoans (*Fenestella*), and trilobites (especially *Phacops* and the large *Trimerus*) are to be found in the green, brown, and tan shales. Some of the *Fenestella* colonies are several inches across.

A smaller road cut across the highway from this exposure is also productive, as are some boulders on its northern fringes. Unfortunately, the shoulders are narrow at the bases of both cliffs, and the rocks are dangerously unstable in places, so collect carefully.

The next large road cut on the west side of Route 259, about one-quarter mile farther south, is comparatively poor in fossils but worth a visit for its unusually large complete specimens of the brachiopod *Tropidoleptus carinatus*. These may be collected whole, with their shell material intact. At first glance, you may think these are individual shells as they are so thin, but a close look will show that the whole brachiopod is present, with one valve being convex and the other one concave. Also present here are pieces of the giant trilobite *Trimerus;* at least pieces are all I have been able to find.

Fossils I have collected here include:

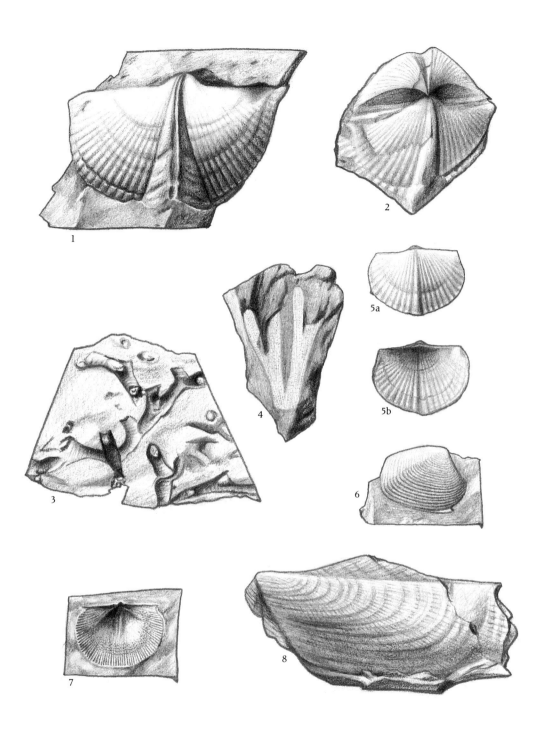

Locality 24. *1, 2,* brachiopods *Spinocyrtia granulosa; 3,* coral *Ceratopora* sp.; *4,* bryozoan *Ptilodictya* sp.; *5a, 5b,* brachiopod *Tropidoleptus carinatus; 6,* pelecypod *Cypricardella* sp.; *7,* brachiopod *Devonochonetes coronatus; 8,* pelecypod *Ptychopteria* sp.

Brachiopods: *Cyrtina hamiltonensis; Delthyris* sp.; *Devonochonetes coronatus; Douvillina* sp.; *Elita (Elytha) fimbriata; Lingula* sp.; *Mucrospirifer mucronatus; Spinatrypa* sp.; *Spinocyrtia granulosa; Tropidoleptus carinatus*

Bryozoans: *Fenestella* sp.; *Ptilodictya* sp.

Cephalopods: straight-shelled nautiloids

Corals: *Amplexus hamiltoniae; Ceratopora* sp.; *Pleurodictyum* sp.; *Stereolasma rectum; Trachypora* sp.

Gastropods: *Loxonema hamiltoniae*

Pelecypods: *Cypricardella* sp.; *Modiomorpha* sp.; *Nuculoidea* sp.; *Palaeoneilo emarginata; Ptychopteria* sp.

Trilobites: *Dechenella* sp.; *Greenops boothi; Phacops rana; Trimerus (Dipleura) dekayi*

Locality 25 in the Town of Lost City, Hardy County, West Virginia

Mahantango formation, middle Devonian period

Locality 25 is a road cut in the town of Lost City on the east side of Route 259, directly across the road from the Friendly Tavern and immediately north of the intersection with Cove Run Road. The wide shoulder just north of the tavern parking area across the road from the outcrop is good for parking, but be sure not to be an obstacle to customers. Local traffic can be moderately heavy at times—by rural standards.

Collecting conditions are not ideal here—the cliff is steep, close to the road, and in a populated area—but fossil preservation is exceptionally good, and a wide variety of species may be found.

Nearly all of the fossils come from a thick layer of shale, grading into a fine-grained sandstone, which first appears at the top of the middle of the road cut and slopes down to ground level at its southern margin. The fissile shales that underlie this layer are quickly weathered into loose chips and slivers, making it very difficult to climb and even more difficult to keep one's footing once the rich layer has been reached. In addition, large chunks of the fossiliferous rock come crashing down at the slightest provocation, which is a hazard to traffic on the highway as well as to fossil hunters. For these reasons, I recommend concentrating on the abundant rockfall, starting directly across the road from the tavern to the southern limit of the outcrop.

Among our most interesting finds are the bottom half of a large *Arthroacantha* crinoid calyx, or crown, sections of a large (over one-inch diameter) straight-shelled *Spyroceras* nautiloid, and a complete, partially pyritized *Phacops rana* trilobite. Many of the trilobite fossils from this locality are covered with minute, sparkling crystals of an unidentified mineral. Besides *Phacops rana,* examples of *Greenops boothi* and *Trimerus dekayi* also occur. Fragments of the latter species give the impression of being covered with short, blunt whiskers; but remember that these fossils are impressions left behind in the rock by shells that have been dissolved away, so the whiskers actually represent pits on the underside of the exoskeleton.

Many of the best-preserved specimens are found on or in extremely hard, cylindrical or lozenge-shaped concretions that occur within the shale layers. These should not be attacked too vigorously with a hammer—and not at all unless you're wearing goggles, gloves, and heavy clothing—because they tend to throw off razor sharp slivers of rock at high speed when and if they decide to break.

I wish there was some way to move this outcrop away from the road, so I could bring down a few tons of fresh rock from the "good" layer; the fossils here are really beautiful, even for the Mahantango formation, which is so productive in the area. But conditions insist that I wait for the treasure to fall, piece by piece.

I have collected the following fossils at this locality:

Brachiopods: *Ambocoelia umbonata; Douvillina* sp.; *Leiorhynchus limitare; Lingula* sp.; *Mucrospirifer mucronatus; Rhipidomella penelope; Spinatrypa* sp.; *Spinocyrtia granulosa; Protoleptostrophia (Stropheodonta) perplana; Tropidoleptus carinatus*

Bryozoans: *Fenestella* sp.

Cephalopods: *Spyroceras crotalum* (straight-shelled nautiloid)

Corals: *Heterophrentis* sp.; *Trachypora* sp.

Crinoids: stems, segments, crowns; *Arthroacantha* sp.

Gastropods: *Platyceras* sp.

Pelecypods: *Cypricardella* sp.; *Lyriopecten orbiculatus*

Trilobites: *Dechenella* sp.; *Greenops boothi; Phacops rana; Trimerus (Dipleura) dekayi*

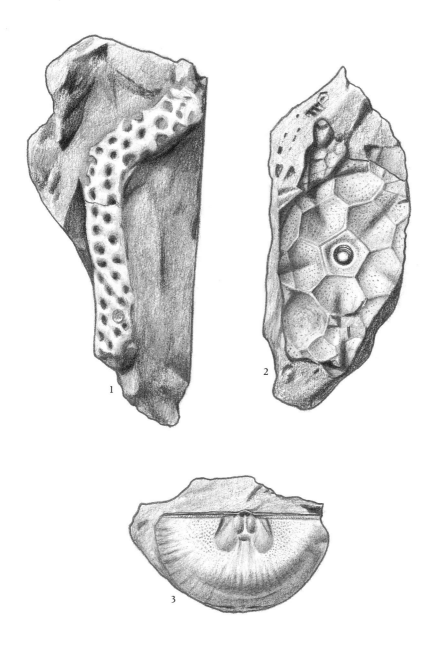

Locality 25. *1,* coral *Trachypora* sp.; *2,* crinoid *Arthroacantha* sp. (external mold of the underside of a calyx); *3,* brachiopod *Douvillina* sp. (internal mold). (Fossils shown 1.5x actual size.)

Locality 26 south of Moorefield, Hardy County, West Virginia

Mahantango formation, middle Devonian period

Locality 26 is a road cut and roadside quarry on the east side of Route 7 (South Branch Road), 6 miles south of its intersection with Route 55 in the city of Moorefield, West Virginia, and 1.3 miles north of the bridge over the South Branch of the Potomac River. The shoulders opposite the road cut and on both sides of the road at the roadside quarry are adequate for parking.

Fossils may be collected from the rocks exposed in the road cut and the roadside quarry to its south. My best luck was in the fairly thick, dark gray (weathering to tan) limestone layers, especially in the northern part of the road cut. This rock is very hard and difficult to break open but contains some excellent brachiopods (*Spinatrypa, Mucrospirifer*) and trilobites (*Trimerus, Phacops*). The trilobite fossils often have their exoskeletons preserved as black calcite, which retains the delicate surface ornamentation. Some of these fossils also occur in the thin-bedded shales surrounding the limestone layers, from which they can be retrieved more easily; these specimens are, however, more likely to be crushed or flattened.

Keep an eye open for quartz crystals. I found a few chunks of rock that had solitary half-inch-long clear crystals cemented to them. These turned up just to the north of an immense boulder that sits on the hillside on the north end of the road cut. Notice the many fossils revealed on its surface by weathering, especially the brachiopods *Cyrtina* and *Ambocoelia*.

The shales exposed in the southern part of the road cut and, to a lesser extent, in the roadside quarry also contain many fossils, including corals (*Pleurodictyum, Amplexus*), crinoids, and brachiopods. But the greatest promise of this locality is only realized through patience and careful collecting, since the best fossils are locked in the most stubborn layers. I have found the following fossils here:

Brachiopods: *Ambocoelia umbonata*; *Athyris spiriferoides*; *Cyrtina hamiltonensis*; *Delthyris* sp.; *Mucrospirifer mucronatus*; *Spinatrypa* sp.; *Spinocyrtia granulosa*
Bryozoans: *Fenestella* sp.
Cephalopods: *Michelinoceras* sp. (nautiloid)
Corals: *Amplexus hamiltoniae* (solitary horn coral); *Pleurodictyum* sp. (colonial)
Crinoids: stems and segments
Gastropods: *Loxonema hamiltoniae*
Pelecypods: *Modiomorpha concentrica*
Trilobites: *Phacops rana*; *Trimerus (Dipleura) dekayi*

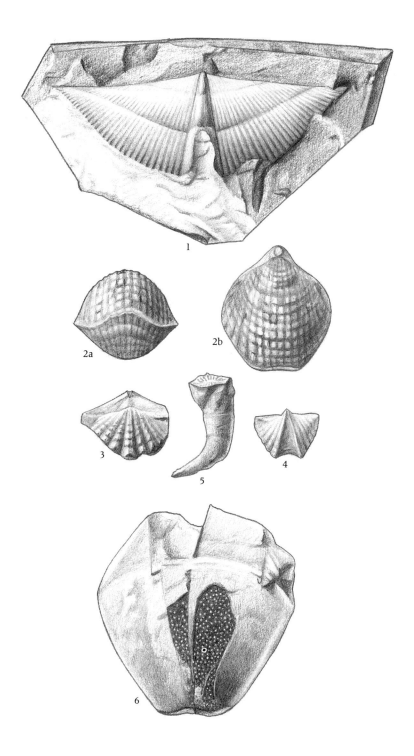

Locality 26. *1*, brachiopod *Mucrospirifer mucronatus*; *2a, 2b*, brachiopod *Spinatrypa* sp.; *3, 4*, brachiopods *Delthyris* sp.; *5*, horn coral *Amplexus hamiltoniae*; *6*, trilobite *Trimerus dekayi* (tail). (Specimens shown 1.2x actual size.)

Locality 27 — north of Rio, Hampshire County, West Virginia

Mahantango formation, middle Devonian period

Locality 27 is a large road cut on the west side of Route 11, 4 miles north of the junction with Route 29 in the village of Rio, West Virginia, and 1.5 miles south of the village of Delray. Fossils also may be found in the numerous boulders along both sides of the road and along the bank of the North River at the bottom of the hill opposite the exposure.

The shoulders are very narrow here. There is one wide area near some large boulders on the east side of the road, immediately opposite the highest parts of the road cut. Parking is also possible along the grassy shoulder just south of the locality. A barbed wire fence, which is hard to see in places, marks the boundary of a private farm that begins on the hillside, about halfway between the road and the river. If you want to collect below this line, ask permission from the landowner first.

An unusually wide variety of fossils may be found at this vast locality. Specimens are easiest to collect from the large boulders of brown, gray, and green shale that abound here, especially around the parking area just north of the middle of the road cut. As at the Mahantango formation localities east of Wardensville, West Virginia (locality 23), and in Lost City, West Virginia (locality 25), many of the best fossils occur in or on the surface of very hard limestone nodules. But the shales that contain them are also highly fossiliferous. Complete brachiopods (*Mucrospirifer, Ambocoelia, Rhipidomella, Athyris*), crinoid stems, gastropods (*Loxonema*), and occasional trilobites (*Greenops, Phacops, Trimerus*) may be found in both types of rock.

Among the more interesting specimens we found were a complete *Greenops*

Locality 27 *(opposite)*. *1*, trilobite *Greenops boothi* (the front of the head is missing); *2*, trilobite *Phacops rana* (enrolled specimen, 1.1x actual size); *3*, trilobite *Trimerus dekayi* (partial head); *4*, brachiopod *Rhipidomella penelope*; *5*, brachiopod *Mucrospirifer mucronatus*; *6, 7*, horn corals *Heterophrentis* sp. (7 is a cut and polished cross section); *8*, brachiopod *Desquamatia reticularis*; *9*, brachiopod *Athyris spiriferoides*; *10*, gastropod *Platyceras* sp.; *11, 12*, gastropods *Loxonema delphicola* (encrusted with bryozoans); *13*, brachiopod *Lingula* sp.; *14a, 14b*, brachiopod *Ambocoelia umbonata* (two views of a specimen shown 1.2x actual size); *15*, gastropod *Bembexia* sp.; *16*, pelecypod *Orthonota undulata*; *17*, pelecypod *Phestia* (?) sp.; *18*, bryozoan *Fenestella* sp.; *19*, nautiloid *Michelinoceras* sp.; *20*, crinoid stem. (Specimens other than *2* and *14a–14b* shown .6x actual size.)

Locality 27, Rio, West Virginia

trilobite fossil with minute black crystals on its surface and a nearly complete head of the trilobite *Trimerus,* which is so large it must have come from an eight-inch-long individual. We also collected several specimens of the gastropod *Loxonema,* which were completely covered by encrusting bryozoan colonies.

The small boulders that line the eastern shoulder of the road near the southern margin of the road cut contain quite different fossils. These calcareous sandstones are riddled with bryozoans (*Fenestella*), large brachiopods (*Spinocyrtia*), and especially horn corals (*Heterophrentis, Zaphrentis*). These fossils, which are preserved in white calcite, are virtually impossible to extract whole from fresh material. The best specimens are obtained from boulders that have been weathered just enough to loosen the fossils, but not so much that surface details have been lost through dissolution of the calcite. The bryozoans and smaller specimens are best appreciated as external molds in more weathered samples.

With my rock saw, I cut some of the horn corals in half, perpendicular to their lengths. After polishing, the beautifully preserved internal structure of these fossils was clearly visible.

These are the fossils I have collected from this locality:

Brachiopods: *Ambocoelia umbonata; Athyris spiriferoides; Desquamatia (Atrypa) reticularis; Devonochonetes* sp.*; Elita (Elytha) fimbriata; Lindstromella aspidium; Lingula* sp.*; Mucrospirifer mucronatus; Rhipidomella penelope; Spinatrypa* sp.*; Spinocyrtia granulosa; Tropidoleptus carinatus*

Bryozoans: unidentified encrusting species; *Fenestella* sp.

Cephalopods: *Agoniatites* (?) sp. (coiled ammonoid); *Michelinoceras* sp. (straight-shelled nautiloid); *Spyroceras* sp. (straight-shelled nautiloid)

Corals: *Heterophrentis* sp. (horn coral); *Zaphrentis* sp. (horn coral)

Crinoids: stems and segments

Gastropods: *Bembexia (Pleurotomaria)* sp.; *Loxonema delphicola; Loxonema hamiltoniae; Platyceras* sp.

Pelecypods: *Actinopteria decussata; Cypricardella* sp.*; Grammysiodea alveata; Modiomorpha concentrica; Nuculoidea* sp.*; Orthonota undulata; Phestia* (?) sp.

Trilobites: *Greenops boothi; Phacops rana; Trimerus (Dipleura) dekayi*

Locality 28   near Gainesboro, Frederick County, Virginia

Mahantango formation, middle Devonian period

Locality 28 consists of road cuts extending for about 0.5 of a mile along both the southbound and northbound lanes of Route 522, 2 miles northwest of the village of Gainesboro and 9 miles northwest of the junction of routes 37 and 522 in Winchester, Virginia. The locality is situated 10 miles southeast of the West Virginia State line. The most productive exposures are located just south and about 0.3 of a mile north of the bridge over Isaacs Creek.

Parking along the highway shoulders is not recommended. Traffic is fairly heavy and fast moving, and the shoulders are narrow. The best parking is at the wide entrance to a private drive along the southbound lanes about 0.3 of a mile north of the bridge, at the picnic area in the median strip immediately south of the bridge, and a bit further south on the wide shoulder on the west side of the southbound lanes.

The Mahantango formation was formerly called the Hamilton formation in this part of the country. The easily collected specimens at most localities are the molds and casts left behind in shale or sandstone by the dissolution of the calcium carbonate shell material, and this is true of most of the fossils found in the green, tan, brown, and gray rocks at this locality.

This has been a magical place for us. Besides being the first good source of trilobite fossils we found, it has produced beautiful quartz crystals, encounters with a variety of wild animals, some good fishing, and, on one occasion, even paper money scattered along the road. It seems strange to have had so many memorable experiences in the median strip of a busy highway.

The best quartz crystals occur between layers of brown sandstone at the northernmost exposure in the median strip along the southbound side of the road. Individual crystals seldom exceed a half inch in length, but clusters of the well-formed, clear specimens can be spectacular. Be careful collecting here, though, as the cliff is very close to the road, the lanes are narrow, and the fast-moving traffic seems intent on blowing you off your feet.

More relaxed collecting can be done about 200 yards to the south, directly opposite the private drive and wide shoulder on the boulder-strewn hillside in the median strip. These rocks are not highly fossiliferous, but a diligent search should turn up trilobite fossils (*Phacops, Trimerus*). Enrolled specimens of *Phacops rana* are the most common finds, many of which are encrusted with minute sparkling quartz crystals. This was the site of my first decent trilobite finds—and of one of my most painful memories. On an early visit, and with a severe case of trilobite fever, I found an extremely rare but unspectacular-looking fish fossil. When some careless handling broke off the tail, I threw it away and have been kicking myself ever since.

More abundant (but less elegant) specimens may be found on the very large road cut immediately across the northbound lanes of the highway. The shales here are mostly tan and gray and contain many crinoid specimens. I found a couple of examples in which the fine, featherlike pinnules, or branches, are still attached to the arms of the crown. Trilobites are also found in the outcrops on the median and on the west side of the southbound lanes immediately south of Isaacs Creek. Those on the median tend to be distorted, however.

By the way, Isaacs Creek is inhabited by an industrious colony of beavers, just downstream of the northbound bridge. Their antics can be a nice diversion after a hard day at the trilobite mine.

The fossils I have collected are:

Brachiopods: *Devonochonetes* sp.; *Mediospirifer audacula; Mucrospirifer mucronatus; Spinocyrtia granulosa; Tropidoleptus carinatus*
Bryozoans: *Fenestella* sp.
Cephalopods: *Michelinoceras* sp. (straight-shelled nautiloid)
Corals: *Heterophrentis* sp. (horn coral)
Crinoids: stems, segments, partial crowns
Gastropods: *Bembexia (Pleurotomaria) ella* (low spired); *Loxonema delphicola* (high spired)
Pelecypods: *Cornellites flabella; Grammysia bisulcata; Nuculites oblongatus; Orthonota undulata; Pterinea* sp.
Trilobites: *Phacops rana; Trimerus (Dipleura) dekayi*

Locality 29 near Hedgesville, Berkeley County, West Virginia

Mahantango formation, middle Devonian period

Locality 29 is a road cut on the northeast side of Route 9, 0.5 of a mile northwest of the Hedgesville town limits. The exposure is 5.9 miles northwest of Interstate Route 81 exit 16W, which is just north of the city of Martinsburg, West Virginia.

There are wide shoulders about 0.2 of a mile north of the most prominent exposures. Traffic can be heavy, and the cliff is very close to the road—so collect with caution.

My first visit to this locality was made in 1966, when I was fourteen years old. It was to be a quick stop, so I ran directly to a boulder at the base of the cliff. Smack in the middle was a large cast of the brachiopod *Spinocyrtia granulosa* (shown in the illustration). I took the whole rock home and found many more specimens within it. Without benefit of a geologic map of the area, I mistakenly identified the fossils as being from the Ordovician Martinsburg formation (which *was* named for the nearby city of Martinsburg, after all). I spent the next year or so gradually extricating myself from my blunder.

Actually, my error was not entirely ridiculous. Both the Mahantango and Martinsburg formations contain brown and green shales, as well as fossiliferous sandstones in their uppermost beds. Both were formed during episodes of mountain building, in similar shallow-water environments. There are even superficial similarities in their fossil contents. But the consequences of a fifty-million-year difference in age finally dawned on me; I just couldn't explain away so many spiriferid brachiopods, which are so typical of Devonian rocks.

I made several more visits during the next twenty years but until quite recently was not able to locate the layers from which my first chunk of fossiliferous shale was derived. At the bend in the road, at the southeastern end of the locality—where the rock cliffs jut out into the highway and are closest to traffic—several layers of calcareous shale, two or three inches thick, occur at knee to chest height. Slabs of this material are not easy to excavate from the cliff face, but it's worth a try, as they are packed with a wide variety of fossils. Spiriferid brachiopods (*Mucrospirifer, Spinocyrtia*) are the most common fossils, but keep an eye open for bryozoans (*Fenestella*) and the familiar trilobite *Phacops rana*.

Phacops rana is probably the best known American trilobite. Its distinctive froglike head and bulging eyes (*rana* means frog) peek from rocks throughout the local mountains in several formations. Perhaps if I had found more than a couple of incomplete tails on my initial visit, I wouldn't have caused myself so much trouble.

The fossils I have collected at this locality are:

Brachiopods: *Devonochonetes coronatus; Leiorhynchus limitare; Mediospirifer audacula; Mucrospirifer mucronatus; Protoleptostrophia (Stropheodonta) perplana; Rhipidomella penelope; Spinocyrtia granulosa; Tropidoleptus carinatus*
Bryozoans: *Fenestella* sp.; unidentified branching species
Corals: *Heterophrentis* sp.
Crinoids: stems and segments
Gastropods: *Bembexia (Pleurotomaria)* sp.
Pelecypods: *Actinopteria decussata; Cornellites flabella; Modiomorpha concentrica*
Pteropods: needle-shaped gastropodlike mollusks
Trilobites: *Greenops boothi; Phacops rana*

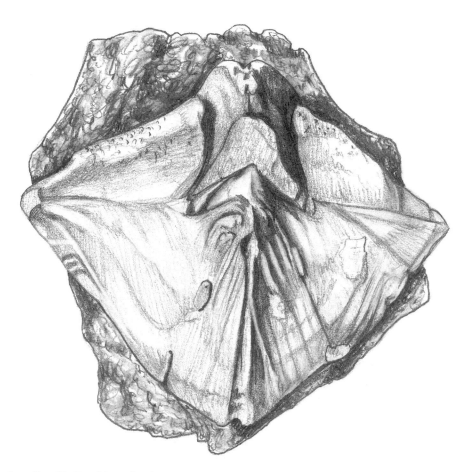

Locality 29. Brachiopod *Spinocyrtia granulosa* (an internal mold, also known as a cast or steinkern). (Specimen shown 2x actual size.)

Locality 30 in Massanutten Mountain, Shenandoah County, Virginia

Mahantango formation, middle Devonian period

Locality 30 is a road cut at the intersection of Routes 678 and 771, 4 miles north of the village of Detrick, Virginia. The exposure is approximately 10 miles south of the junction of Route 619 with Route 55 at the village of Waterlick (5 miles east of Strasburg). From Route 55, turn south on Route 619, bear right on Route 678 after 1 mile or so, then continue about 9 miles to the locality. Traffic is very light; parking space is available on the northern shoulder of Route 771.

Massanutten Mountain is a forty-mile-long, canoe-shaped system of ridges in the midst of the Shenandoah Valley. Its northern interior has been eroded by Passage Creek, which rushes by on the east side of Route 678 at the locality. The best collecting here is at the road junction itself, although fossils may also be found in the road cuts and scattered boulders along both roads and the stream.

The ravages of mountain building may be seen in many of the specimens. In particular, the abundant fossils of the wide brachiopod *Mucrospirifer mucronatus* are generally distorted or crushed due to the pressures that these dark green and brown shales and siltstones were subjected to. These effects are less apparent in smaller specimens such as corals (*Pleurodictyum*) and tentaculitids. Some rocks contain large numbers of the latter, stacked on top of each other.

I first collected here in 1967 but have never stayed long enough to investigate it fully. I am sure that many more species of fossils may be found than are listed below.

Brachiopods: *Cupularostrum (Camarotoechia) congregatum; Devonochonetes coronatus; Mucrospirifer mucronatus; Tropidoleptus carinatus*
Corals: *Pleurodictyum* sp.
Tentaculitids: *Tentaculites scalariformes*
Trilobites: *Greenops boothi*

Locality 31 near Edinburg, Shenandoah County, Virginia

Mahantango formation, middle Devonian period

Locality 31 is a road cut on the north side of Route 675, just east of the intersection with Route 717, 8.6 miles west of the town of Edinburg, Virginia. The locality is 3.6 miles south of the West Virginia State line and 4.2 miles south of locality 3. From Interstate Route 81, take exit 71 to Route 675 West and proceed 8 miles. (Follow the signs for Route 675 carefully, as it joins Route 42 for a short distance.) There is ample parking on both sides of the road.

The best collecting here is in the somewhat unstable layers that slope toward the road in the middle of the exposure. The gray and green shales contain many brachiopods (*Spinocyrtia, Delthyris*), crinoids, and corals, especially examples of the horn coral *Heterophrentis*. The shale is frequently stained by yellow, red, and purple mineral residues, which often enhance fossil specimens. Collecting is safest from the scree and from the debris that has been piled up on the western fringe of the road cut and across the road.

As at many Mahantango formation localities, the abundant horn corals are difficult to extract from fresh rocks. The corals themselves are composed of brittle calcite, which breaks more readily than the shale matrix. The best specimens I found were external molds of the tops of the corals, which preserve details of their intricate structures. The cuplike opening of each coral is represented by a dome-shaped mold. Similar fossils may be collected from road cuts to the north along Route 675, which turns sharply north just west of the locality. I found excellent brachiopod fossils on the west side of the road near a small parking area four-tenths of a mile to the north. Here are the fossils I have collected:

Brachiopods: *Athyris spiriferoides; Delthyris* sp.; *Desquamatia (Atrypa) reticularis; Devonochonetes coronatus; Elita (Elytha) fimbriata; Mediospirifer audacula; Mucrospirifer mucronatus; Protoleptostrophia (Stropheodonta) perplana; Rhipidomella* sp.; *Spinatrypa* sp.; *Spinocyrtia granulosa; Tropidoleptus carinatus*

Bryozoans: *Fenestella* sp.

Cephalopods: fragments of straight-shelled nautiloids

Corals: *Heterophrentis* sp.; *Trachypora* sp.

Crinoids: stems, segments, parts of crowns

Gastropods: *Bembexia (Pleurotomaria)* sp.

Pelecypods: *Actinopteria decussata; Cornellites flabella*

Trilobites: *Phacops rana; Trimerus (Dipleura) dekayi*

Locality 32 near Danville, Allegany County, Maryland

Mahantango formation, middle Devonian period

Locality 32 exposures border the parking lot at Dale's Pit Stop on the west side of U.S. Route 220, 1.5 miles south of the village of Danville and 9.5 miles south of Cresaptown, Maryland. The locality is approximately 14.5 miles south of Cumberland and 4.5 miles north of Keyser, West Virginia. The collecting area is on private property so *ask for permission to collect from the owner of the store.* Parking is plentiful.

The rocks exposed here contain an abundance of corals, both colonial tabulate forms (*Favosites*) and solitary horn corals (*Heliophyllum*). The specimens are so thick in some places that they compose much of the rock, stacked on top of each other in reeflike mounds, but preservation is generally poor. Collecting is easy, however, since the debris at the base of the cut includes many loose fossils.

A close examination of the growth lines on better preserved specimens of horn corals like *Heliophyllum* once led to a startling discovery. Scientists who counted these lines, which are analogous to tree rings but record daily cycles as well as annual ones, found that the Earth spun more rapidly on its axis in the Devonian period—and a year consisted of about four hundred days!

Before helping yourself to the fossils here, ask for permission in the store. We visited this unusual site in the evening, and the owner seemed pleased to show off his fossil locality. He invited us to take samples and even offered to lend us flashlights so we could continue to collect when the light grew dim. But I would suggest taking just a few loose specimens and leaving the cliff intact. This is one place where the interest of the whole exceeds the sum of its parts; the individual fossils are not very impressive, but the spectacle of so many marine coral fossils on one mountain roadside is very much so. I went away with only

Corals: *Favosites* sp.; *Heliophyllum* sp.

Locality 33   near Flintstone, Allegany County, Maryland

Chemung formation, late Devonian period

Locality 33 is a large road cut on the north side of U.S. Route 40-48, 4.5 miles east of the Chaneysville Road exit to Flintstone, Maryland, and 2.4 miles west of Route 40-48 exit number 64. The locality is 16 miles east of the city of Cumberland and 11 miles east of locality 8. Route 40-48 is a busy highway, so collect with caution. The shoulder is sufficiently wide for parking.

The fine-grained sandstones and shales exposed here are fossiliferous only in certain areas. Our most productive collecting was done near the middle of the road cut. Preservation is only moderately good, though exceptions occur in widely scattered layers of tan shale and in some purple sandstones. Many of the well-preserved fossils in the latter are coated with yellow iron oxide, which makes them very nice to look at.

Gastropods, brachiopods, and crinoid fragments are the most common finds. Unusual specimens include rare *Phacops rana* trilobites (we found one) and pieces of plant material that may have floated out to sea from the coastal forests that once grew nearby. All of the fossils we found consisted of molds and casts.

This locality is being altered by road-building activities in the area. The illustration shows the exposure as it was in early 1989. Since that time, it has been terraced into a regular series of fairly low steps cut into the bedrock. This change, although it makes my picture less useful for recognition purposes, is a boon to collectors, as it has exposed a great deal of fresh fossil-bearing rock. The steps also make the exposure much easier to climb.

Fossils I have collected here are:

Brachiopods: *Cyrtospirifer disjunctus*; *Leiorhynchus* sp.; *Mucrospirifer* sp.; *Productella* sp.; *Productella (?)* sp.; *Ptychomaletoechia* sp.; *Tropidoleptus carinatus*; *Tylothyris (Delthyris) mesocostalis*; *Tylothyris* sp.
Crinoids: stems and segments
Gastropods: *Cyclonema* sp.; *Holopea* sp.; *Murchisonia* sp.; *Platyceras* sp.
Pelecypods: *Aviculopecten cancellatus*
Plants: fragments
Trilobites: *Phacops rana*

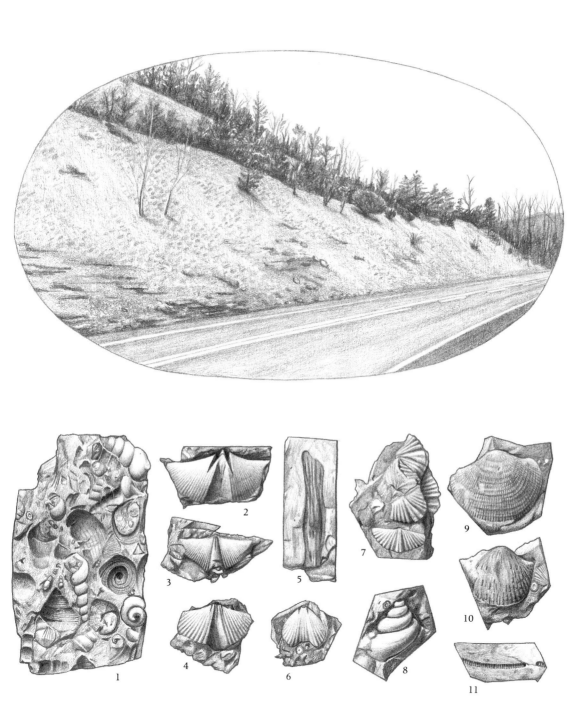

Locality 33. *1,* gastropods *Murchisonia* sp. (high spired) and *Cyclonema* sp. (low spired); *2, 3,* brachiopods *Mucrospirifer* sp.; *4,* brachiopod *Tylothyris mesacostalis; 5,* plant fragment; *6,* brachiopod *Ptychomaletoechia* sp.; *7,* brachiopods *Tylothyris* sp.; *8,* gastropod *Cyclonema* sp. (internal mold); *9,* brachiopod *Productella* (?) sp.; *10,* brachiopod *Productella* sp. (different species from *9*); *11,* crinoid stem section. (Fossils shown .6x actual size.)

Locality 34 between Keenan and Gap Mills, Monroe County, West Virginia

Greenbrier group, late Mississippian period

Locality 34 consists of road cuts on both sides of Route 3, 2.4 miles east of Keenan and 2.1 miles west of Gap Mills. The exposures are 0.2 of a mile east of Doss Shaver Road. The reddish sandy banks on the east end of the locality are the most productive. The shoulders are wide enough for parking at either end of the road cut. Traffic is light.

The Greenbrier group includes many limestones deposited in the widespread continental seas of the late Mississippian period. Two very different kinds of collecting may be done at this locality. The western part of the road cut consists of exposures of thick-bedded, dark gray limestone. These contain coral fossils (*Acrocyathus*), which are difficult to extract but worth taking a look at. In contrast are the easy pickings in the reddish sandy soils on the road banks to the east. These are littered with yellowish chert nodules and loose fossils, which were left behind when the limestones that once contained them weathered away.

The most obvious species of fossil found here is the round brachiopod *Inflatia inflata*. Specimens average about an inch across and are very common. A close look at the surface of the soil banks reveals many smaller fossils, including horn corals (*Triplophyllum*) and abundant crinoid and blastoid stems and segments.

Don't hesitate to get on your hands and knees and plunge into the soil, sifting for small fossils. I discovered that the dirt brushes off very easily (at least when dry) and was rewarded for my digging when I found a small, complete blastoid crown (*Pentremites*), which I would never have seen from a distance.

The reason the fossils have survived while the limestone that once contained them has been reduced to sand and dirt is that they are made of chert, a form of quartz, which is highly resistant to most kinds of weathering. Many of the chert nodules have fossils standing out on their surfaces. I do not recommend trying to completely separate these fossils from their matrices, because they seldom break free cleanly and the light gray color of the jagged, freshly broken chert does not complement the orange or yellow fossils well. I advise trimming off the excess matrix with a rock saw.

Fossils added to my collection include:

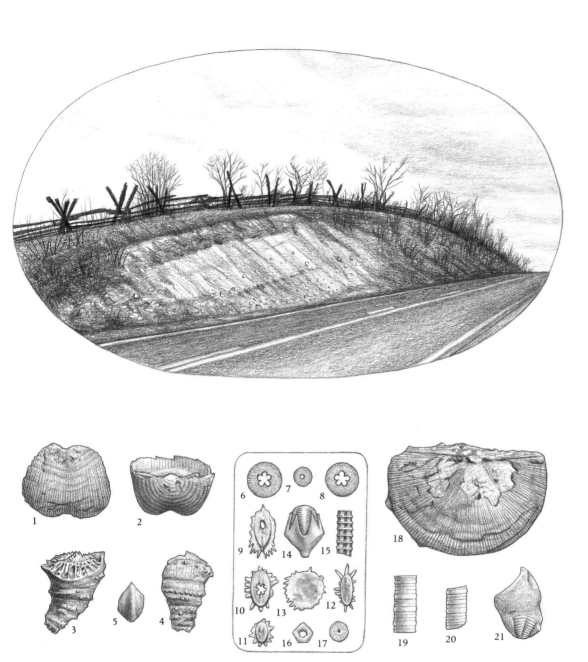

Locality 34. *1, 2,* brachiopods *Inflatia inflata; 3, 4,* horn corals *Triplophyllum* sp.; *5,* gastropod; *6–8,* crinoid or blastoid segments; *9–12,* crinoid stem segments; *13,* plate from crinoid calyx; *14,* blastoid *Pentremites* sp. (calyx); *15–17,* crinoid or blastoid stem fragments; *18,* brachiopod *Orthotetes* sp.; *19, 20,* crinoid or blastoid stem sections; *21,* trilobite *Kaskia* (?) sp. (tail). (Specimens *6–17* shown 1.5 actual size, others shown .7x actual size.)

Blastoids: *Pentremites* sp. (stems, segments, crown, or calyx)
Brachiopods: *Composita trinuclea; Inflatia (Productus) inflata; Orthotetes* sp.; *Spirifer* sp.
Bryozoans: *Fenestella* sp.
Corals: *Acrocyathus (Lithostrotionella)* sp.; *Triplophyllum* sp.
Crinoids: stems, segments, plates from crowns
Gastropods: unidentified species
Trilobites: *Kaskia* (?) sp. (tail, or pygidium)

Locality 35 in Locust Creek near Hillsboro, Pocahontas County, West Virginia

Greenbrier group, late Mississippian period

Locality 35 fossils occur in chunks of chert in the stream gravel of Locust Creek, above and below the stone bridge on Locust Creek Road, 1.5 miles southeast of Route 219. From Hillsboro, take Route 219 southwest for 2 miles, then turn left on Locust Creek Road. There is a wide parking area north of the bridge. Private farmland borders the stream, so limit collecting to the streambed itself, or ask for permission to explore elsewhere.

Nearly all of the fossils at this locality are examples of the colonial horn coral *Acrocyathus*, preserved in pieces of green, tan, and especially blue chert. Many specimens are translucent, so the internal structure may be seen in small samples or in thin slices cut with a rock saw and polished. Many of the greenish chunks turn out to be sky blue inside when cut in this way.

Except when the water is very low, the only way to collect is to wade into the stream itself and look for green or blue stones on the bottom. We found so much of this material that we soon became very selective in what we kept. Occasionally, a specimen will turn up that preserves the external structure of the colony on its surface, but most pieces have been broken and smoothed by the stream action.

At one time, this chert, which is made of weather-resistant quartz, was contained in layers of limestone. When the limestone was dissolved away, the chert was left behind to litter the landscape—and to beautify fossil collections. My own was enriched with

Brachiopods: unidentified species
Corals: *Acrocyathus (Lithostrotionella) floriformis*
Crinoids: segments

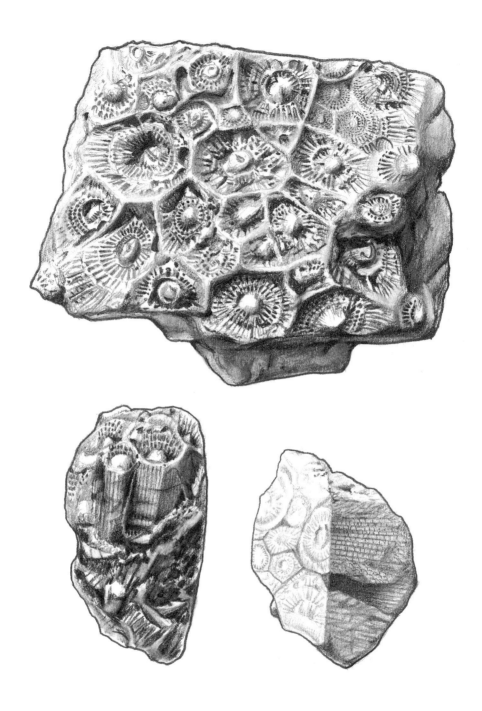

Locality 35. Corals *Acrocyathus (Lithostrotionella) floriformis*. Specimen on lower right has been cut and polished on its left side.

Locality 36 — east of Uniontown, Fayette County, Pennsylvania

Mauch Chunk formation, late Mississippian period

Locality 36 is a large quarry just north of U.S. Route 40 at the base of Chestnut Ridge, 1.2 miles east of its summit, 8 miles east of Uniontown, Pennsylvania. The locality is approximately 1 mile west of the village of Chalk Hill.

The grounds surrounded by this inactive quarry are now used by the Pennsylvania Department of Transportation for gravel and equipment storage. Park on the wide shoulder on the north side of Route 40, well clear of the entrance. The best collecting is to the west (uphill) of the storage area.

These limestone layers of the Mauch Chunk formation, known as the Wymps Gap limestone member, were formerly referred to as the Greenbrier limestone. The medium to light gray rocks (which include some shales) are rich in well-preserved, easily collected fossils.

About twenty feet above ground level in the northwestern corner of the quarry, there is a narrow recess into the cliff face. This represents the efforts of many collectors who have located and excavated a particularly rich layer here. Rocks from this and many other layers around the western wall of the quarry are packed with brachiopods as well as sections of crinoid stems and occasional specimens of the trilobite *Kaskia*. These fossils are most easily collected from the abundant scree, or loose rocks, in this area.

Many of the same fossils are also present in the rocks that litter the floor of the quarry and hillside just south of the above exposures, especially in the vicinity of a small outcrop. Many of these rocks are flat slabs whose surfaces are covered by fossils, including the brachiopod *Orthotetes* and the bryozoans *Fenestella* and *Polypora*. These bryozoans are very similar in appearance, both consisting of flat, lacy colonies. A hand lens is needed to tell them apart. In *Polypora*, each thin branch has from three to eight longitudinal rows of holes, known as zooecia, on each side, each of which represents an individual animal. Branches of *Fenestella* colonies have only two rows of zooecia and on one side only.

Also abundant on these rocks are the small, pinlike spines that were originally attached to shells of the brachiopod *Diaphragmus*. A close look at shell specimens reveals where these spines have been broken off.

If you should happen to visit on a hot day, look for a vent, or hole in the ground, near the southern end of the western wall of the quarry. A cool breeze blows out of this from a subterranean cave. This is an excellent locality with a wide variety of beautiful fossils—and air conditioning besides.

I have collected these fossils from this locality:

Brachiopods: *Composita subquadrata; Diaphragmus cestriensis; Linoproductus ovatus; Orthotetes* sp.; *Rhipidomella* sp.; *Spirifer increbescens*
Bryozoans: *Fenestella* sp.; *Polypora* sp.; branching species (twiglike)
Corals: *Aulopora* sp. (encrusting); unidentified horn coral
Crinoids: stems, segments, arms of crowns; *Eupachycrinus* sp.
Gastropods: *Strobeus* sp.
Pelecypods: *Wilkingia* sp.
Trilobites: *Kaskia* sp.

Locality 36 *(opposite)*. *1*, pelecypod *Wilkingia* sp.; *2*, horn coral; *3, 4*, bryozoans *Fenestella* sp. (note corals *Aulopora* sp. on the surface of *4*); *5, 6*, crinoid *Eupachycrinus* sp. (stem sections); *7*, trilobite *Kaskia* sp. (tail); *8*, brachiopod *Linoproductus ovatus*; *9, 10*, brachiopods *Composita subquadrata*; *11, 12*, brachiopods *Spirifer increbescens*; *13, 14*, brachiopods *Diaphragmus cestriensis*; *15*, brachiopod *Orthotetes* sp. (Specimens shown .6x actual size.)

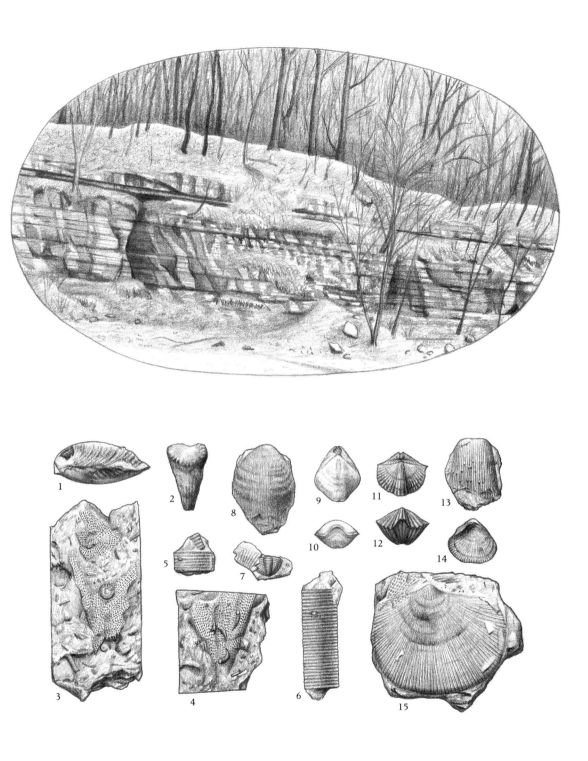

Locality 36, Uniontown, Pennsylvania

Locality 37 along Scenic Route 150 northwest of Marlinton, Pocahontas County, West Virginia

Kanawha formation, middle Pennsylvanian period

Locality 37 is an extensive road cut on the west side of Scenic Route 150 about 12 to 15 miles north of its junction with Route 39 and 2.3 miles south of the intersection with Route 86 at the Williams River bridge. From Marlinton, take Route 39 west approximately 15 miles to the Cranberry Mountain Visitor Center, then right on Route 150, and continue to the locality, which will be on your left.

The shoulder is just wide enough to permit parking on the same side of the road as the road cut. Traffic is very light. This is one of the most remote parts of the eastern United States south of Maine. Camping facilities are available at Day Run campground, about 8 miles away. To get there, take Route 150 north from the locality, turn right on Route 86 and then take a very sharp right where Route 86 ends. The campground is on the left 1 mile or so down this road, and isolated sites occur here and there along the road beyond it.

Many kinds of plant fossils occur in many different types of rock at this locality. Coarse gray sandstones contain large sections of *Calamites* stems and the trunks, cones (*Lepidostrobus*), and roots (*Stigmaria*) of the scale tree *Lepidodendron.* Some pieces of trunk are three or four feet long and twelve inches in diameter. Finer grained sandstones and green, gray, or brown shales yield beautifully preserved leaf fossils (*Neuropteris, Mariopteris*) and more *Calamites* stems and branches. In fact, so many fossils may be found in such widely scattered parts of this half-mile-long road cut, that it is a bit overwhelming in a short visit. So, considering the remoteness of the area, its remarkable natural beauty, and the proximity of excellent campsites, I suggest making your visit a two-or-three-day expedition.

Very roughly speaking, different parts of the locality have produced fossils for me in the following pattern. The rocks in the southern fringe of the exposure are mostly light gray sandstones containing poorly preserved *Lepidodendron* fossils.

Locality 37 *(opposite).* 1, cone of lycopsid *Lepidostrobus* sp.; 2, root of lycopsid *Stigmaria ficoides,* (shown .2x actual size); 3, 4, seed fern leaves *Alethopteris lonchitica;* 5, base of horsetail stem *Calamites* sp.; 6, 7, seed fern seeds *Trigonocarpus* sp.; 8, seed fern leaf *Neuropteris* sp.; 9, lycopsid (scale tree) trunk *Lepidodendron* sp.; 10, seed fern leaf *Mariopteris* sp.; 11, lycopsid trunk section *Lepidodendron* sp. (Specimens other than 2 shown .6x actual size.)

Locality 37, Marlinton, West Virginia 133

Further north, the road turns to the left. At the bend is an abundance of coal (which is a nice addition to your campfire, by the way) and a few seed fern leaves and *Calamites* fossils. Most of the really large specimens of *Lepidodendron* trunks were found just north of this bend in the road in fallen boulders at the base of the cliff. Many were far too heavy to haul away and are probably still there for you to see. The northernmost stretch of the road cut is where my best-preserved fossils came from, especially in the thin-bedded shales that lie in the upper half of the cliff and accumulate in heaps at its base.

While most of the stems and trunks have been flattened by the pressures of great thicknesses of sediment, some of the specimens I found were remarkably unaltered in shape. Sections of horsetails, in particular, were in excellent condition; with a bit of glue, I was able to reconstruct long sections of the branches and trunks.

Scale tree, or lycopsid, trunks bear distinctive, diamond-shaped markings, which were points of leaf attachment in life. The large parts of their root systems, called *Stigmaria,* have simple circular pits with rounded bumps in their centers. These trees reached heights of nearly 150 feet. Along with the *Calamites* horsetails, which grew close to 100 feet high, they dominated the landscape in the Pennsylvanian period. But as beautiful as those forests must have been, they could hardly have surpassed the forest that flourishes here today.

From this locality, I have collected:

Plants: *Alethopteris lonchitica* (seed fern); *Calamites* sp. (horsetail, or sphenopsid); *Lepidodendron* sp. (lycopsid, or scale tree); *Lepidostrobus* sp. (lycopsid cone); *Mariopteris* sp. (seed fern); *Neuropteris* sp. (seed fern); *Stigmaria ficoides* (lycopsid root); *Trigonocarpus* sp. (seed fern seed)

Locality 38   across the Ohio River from Ambridge, Beaver County, Pennsylvania

Mahoning and Brush Creek formations, middle Pennsylvanian period

---

Locality 38 is a large road cut on the west side of Route 51, on the west bank of the Ohio River, immediately opposite the Ambridge-Woodlawn Bridge, which crosses the river from Route 51 to the town of Ambridge. The locality is 1.7 miles north of the junction of routes 51 and 151 and approximately 2 miles south of the town of Aliquippa. It is 6 miles northwest of Locality 39.

Route 51 is a busy four-lane highway. Fortunately, there is a wide shoulder for parking just south of the ramp to the bridge on the west side of the highway. If you approach this locality from the south on Route 51, you will have to continue past it for 2 miles in order to turn around.

Two very different kinds of fossils may be collected here. The exposure immediately above the highway consists of the green and gray shales and sandstones of the Mahoning formation. These rocks contain plant fossils, especially the fissile shales near the north end of the road cut. Well-preserved specimens of *Neuropteris, Pecopteris,* and *Calamites* were found by carefully "paging" through the thin layers of rock. When collecting from this material, have some Elmer's Glue-all handy and some protective containers, as the rocks often split so thinly that they become fragile and endanger the fossils on them.

Very different kinds of fossils are found in the black shales and limestones of the Brush Creek formation, which are exposed on the next level of the terraced road cut. These rocks are also present as abundant rockfall on the first level, especially near its northern end. Most of our collecting has been in the fallen material, as these higher cliffs are difficult to reach—and we were more than satisfied with what we found below. If you choose to climb to the higher cliffs, be prepared for a rough and somewhat dangerous climb.

The Brush Creek specimens are marine in origin and include brachiopods, cephalopods, gastropods, pelecypods, horn corals, and at least two kinds of teeth from primitive sharklike fish, including *Petalodus.* The shiny, bladelike teeth of this genus are reminiscent of the fossil teeth of the shark *Carcharodon megalodon* that we found in the Tertiary deposits of Maryland and Virginia. In fact, the specimen I found has the same greenish-gray enamel and serrated edges as the much younger teeth I found at localities 44 and 45.

On my last visit to this locality, I noticed that many horsetails were growing on the level parts of the road cut. These are modern relatives of the fossil sphenopsid *Calamites*, whose remains may be found here. It seemed as if the past was not so distant after all. But then I calculated that if every year separating the lives of the fossil plants from those of today was measured as a distance of one inch, then the time interval between them would exceed four thousand seven hundred *miles*!

Fossils I have collected from the Mahoning formation include:

Plants: *Calamites* sp. (horsetail, or sphenopsid; *Neuropteris* sp. (seed fern); *Pecopteris* sp. (herbaceous fern)

Fossils I have collected from the Brush Creek formation include:

Amphibians: jaw fragments of labyrinthodonts
Brachiopods: *Derbyia* sp.; *Chonetinella* sp.
Cephalopods: straight-shelled nautiloids
Corals: *Stereostylus* sp.
Gastropods: *Bulimorpha* sp.; *Cymatospira* sp.; *Meekospira* sp.; *Shansiella* sp.; *Strobeus* sp.
Pelecypods: *Nuculopsis* sp.; *Phestia* sp.
Sharklike fish: *Petalodus* sp.; tooth of unidentified species

Locality 38 *(opposite)*. *1,* brachiopod *Derbyia* sp.; *2,* gastropod *Cymatospira* sp.; *3,* brachiopod *Chonetinella* sp.; *4, 5,* pelecypods *Nuculopsis* sp.; *6,* gastropod *Strobeus* sp.; *7,* gastropod *Bulimorpha* sp.; *8,* fern frond *Pecopteris* sp.; *9,* horsetail stem *Calamites* sp.; *10,* jaw fragment of a labyrinthodont amphibian; *11,* tooth of sharklike fish *Petalodus* sp. (right two-thirds of the root of the tooth is missing); *12,* tooth of sharklike fish; *13,* seed fern leaf *Neuropteris* sp.; *14,* pelecypod *Phestia* sp.; *15, 16,* gastropods *Shansiella* sp.; *17,* horn coral *Stereostylus* sp. (Specimens shown .6x actual size.)

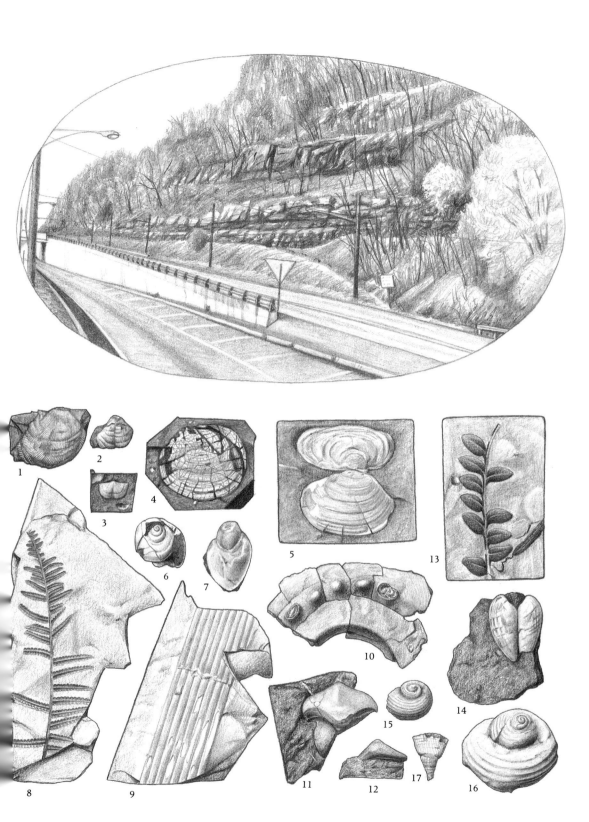

Locality 38, Ambridge, Pennsylvania

Locality 39 across the Ohio River from Sewickley, Allegheny County, Pennsylvania

Brush Creek formation, middle Pennsylvanian period

Locality 39 is a long road cut on the west side of Route 51, directly opposite the bridge that crosses the Ohio River from Route 51 to the town of Sewickley, Pennsylvania. This locality is approximately 6 miles southeast of Locality 38.

A parking area is located just south of the road cut on the west side of the highway. If coming northbound on Route 51, you will have to continue past the locality in order to turn around. Traffic on Route 51 is heavy, but there is a wide level area between the cliffs and the guardrail.

The black shales and limestones of the Brush Creek formation are exposed about ten feet above ground level, immediately opposite the bridge, but collecting is easiest in the many large boulders strewn over the flat area between the road and outcrop for a couple of hundred yards to the north. The fossils in these rocks are similar in kind and appearance to those at locality 38.

Perhaps the most attractive specimens at this locality are the small horn corals *Stereostylus*. They are preserved in white calcite, which contrasts sharply with the very dark matrix. The internal structures of these fossils are easily seen in the cross sections of broken examples. Some brachiopods and pelecypods are also white or silvery and stand out to the eye, but it will take some looking to spot the many darker specimens that occur here.

Fossils collected here include:

Brachiopods: *Reticulatia (Dictyoclostus)* sp.
Bryozoans: *Rhombopora* sp. (branching colonies)
Cephalopods: straight-shelled nautiloids
Corals: *Stereostylus* sp.
Gastropods: *Meekospira* sp. (high-spired); *Shansiella* sp. (low spired)
Pelecypods: *Nuculopsis* sp.; *Allorisma terminale* (3 inches long)

Locality 40 along the G. C. and P. Road near Wheeling, Ohio County, West Virginia

Monongahela group, late Pennsylvanian period

Locality 40 consists of boulders along Long Run in the vicinity of the G. C. and P. Road bridge near the entrance to Whitmar Hills. To get there from Wheeling, take U.S. Route 40 east. Turn north on Route 88, about 2 miles east of downtown, following the signs for Olgebay Park. Proceed about 0.8 of a mile and turn left on the G. C. and P. Road. Follow it for 1.5 miles to the locality. When the stream is on your right, you've gone too far.

There is a wide entrance to a private drive on the southeast side of the road, just before the bridge, with room for parking. Be sure not to block the driveway. A short but steep hill must be negotiated to reach the streambed, so watch your step. Traffic can be heavy on the road at times.

I first visited this locality in 1967 but never exhaust its possibilities, thanks to the frequent floods that renew the boulder supply. The bedrock is best exposed at the bridge but is inaccessible except at very low water levels, so I prefer splitting open the rocks that litter the stream banks from just above the bridge to a hundred yards or so downstream. Several different kinds of rocks are present here, but I have found fossils only in the light to dark gray shales, especially in the very hard, light-shaded boulders that ring when you hit them with a hammer. These resemble slate in texture and hardness.

The most abundant recognizable fossils are the fronds of the fern *Pecopteris*. Each frond contains many leaves, each leaf consisting of numerous small leaflets. I cracked open one slab that held the positive and negative impressions of a branch over two feet long with fifteen fronds still in place. The fossils consist of black carbon films, so the lighter gray the matrix, the more attractive the specimen. On my first visit, I found the mangled remains of a fossilized fish in a very dark gray, almost black shale. It consisted of a shiny mass of scales with a very recognizable fin attached. No other fish has turned up.

Since this shale is very hard and brittle, a chisel or hammer with a flat chisel end is needed for splitting open the boulders. Generally, the rock will break along the fossil, as that is the plane of weakness, but it is difficult to get it to break the way you want it to, so take home the slabs that you know contain fossils and do your close work with small chisels in a place where you can hope to retrieve the slivers that fly away.

A local collector told me he had found a lot of crustacean fossils here. Since he said "a lot" and since I haven't found one in twenty-four years, I think he must

have been fooled by some of the strange-looking root fossils that turn up. They look rather lobsterlike sometimes, with rootlets radiating from the main stock like legs. Still, there is great potential here for unusual fossils. I often find entirely new types that I can't quite identify. An insect fossil wouldn't surprise me, and somewhere, another fish is hiding.

Here are the fossils I have collected at this locality:

Fish: unidentified species

Plants: *Alethopteris zeilleri* (seed fern); *Alethopteris* (?) sp. (seed fern); *Alloiopteris* sp. (herbaceous fern); *Annularia* sp. (leaves of horsetails); *Calamites* sp. (horsetail stems); *Cordaites* sp. (gymnosperm); *Lepidophyllum* sp. (scale tree, or lycopsid, leaves); *Pecopteris arborescens* (herbaceous fern)

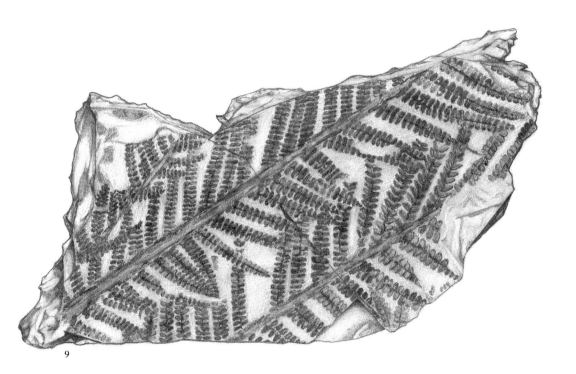

9

Locality 40 *(opposite and above)*. *1*, fern frond *Pecopteris arborescens*; *2*, unidentified plant; *3*, fern frond *Alloiopteris* sp.; *4*, sphenopsid (horsetail) leaves *Annularia* sp.; *5*, lycopsid (scale tree) leaves *Lepidophyllum* sp.; *6*, seed fern leaf *Alethopteris* (?) sp.; *7*, seed fern leaf *Alethopteris zeilleri*; *8*, fern frond *Pecopteris arborescens*; *9*, herbaceous fern fronds *Pecopteris arborescens*. (Specimens *1–8* shown .6x actual size; specimen *9* shown .7x actual size.)

Locality 41 near Bethany, Brooke County, West Virginia

Monongahela group, late Pennsylvanian period

Locality 41 fossils are found in the rockfall at the base of a long road cut on the north side of Route 67, 1.3 miles east of the town of Bethany, and 2 miles east of the junction of routes 88 and 67 on the western edge of town. This exposure is across the road from Buffalo Creek, which is crossed by Route 67 0.8 of a mile west of the locality. The material at the eastern end of the road cut has been the most productive.

The shoulder across the road is sufficiently wide in places for parking, especially opposite the best collecting at the eastern end. Beware of places where loose rocks have been heaped along the road; they make the shoulder seem wider than it really is, and the hillside below is very steep. Traffic is light.

The northern panhandle of West Virginia and adjoining parts of Pennsylvania and Ohio are known for coal beds and the plant fossils associated with them, so my brother Phil and I were somewhat frustrated when several hours of exploring streambeds, road cuts, quarries, and natural exposures produced nothing. As so often happens, our very last stop before turning back turned up precisely what we were looking for.

There is not a great variety of plant species in these light green shales, but preservation is generally excellent, with the fine veins of the leaves distinctly visible in the black carbon films that have replaced the plant material. These carbon layers, which are actually thin sheets of coal, fade and flake off and are often entirely gone from weathered specimens, taking much of the beauty of the fossils with them. So look over the freshest rock you can find.

I suggest concentrating on the rockfall rather than removing slabs from the cliff itself. The most fossiliferous layers are high up in the road cut and are very unstable, especially when the rocks are wet. On several occasions, I have collected from the rubble as chunks of rock rained down all around me.

The best collecting is usually in the finer grained shales with a smooth homogeneous texture. Split the rocks carefully with your hammer or chisel, as they break easily—and usually right through the best specimens. Nearly all the fossils are of the large individual leaves of the seed fern *Neuropteris scheuchzeri*. At least one other species of *Neuropteris* turns up now and then, as well as stems, roots, and the occasional fanlike leaves of the horsetail, known as *Annularia stellata*.

Fossils collected here are:

Plants: *Annularia stellata* (horsetail leaves); *Neuropteris scheuchzeri* (seed fern); *Neuropteris* sp. (seed fern)

Locality 41. *1–3,* seed fern leaves *Neuropteris* sp.; *4,* leafless seed fern stem *Neuropteris scheuchzeri* (those are not thorns); *5,* sphenopsid (horsetail) leaf *Annularia stellata; 6,* seed fern leaf *Neuropteris* sp.; *7,* seed fern leaf *Neuropteris scheuchzeri* (note partial *Annularia stellata* leaf in upper left). (Specimens shown .6x actual size.)

Locality 42 near Delaware City, New Castle County, Delaware

Mount Laurel formation, Cretaceous period

Locality 42 consists of spoil banks on the north shore of the Chesapeake and Delaware Canal. To reach the collecting area from Route 9 in Delaware City, proceed south, turn right onto the road that parallels Route 9 starting immediately past the small bridge over the *small* canal. Follow this road for 0.5 of a mile to its end. Turn left up a small hill and park. Fossils can be found on the extensive sandy area over the small hill on your left. (Similar fossils may be found on the south side of the C & D Canal just east of Route 9, but in less profusion.) The parking area is large and convenient to the deposits.

Fossils are extremely abundant here and easily collected. Innumerable belemnoid specimens and shells of the large extinct oysters *Exogyra* and *Pycnodonte* are scattered over this vast locality. I had to upgrade the contents of my rock bag continuously so it didn't become too heavy to carry.

The Mount Laurel formation does not crop out naturally in this area, and its fossils are accessible only because of the activities of canal builders, who have scooped out the deep channel and piled the spoils on either bank. By far the most abundant fossils are the amber-colored, translucent belemnoids. These bullet-shaped shells gave internal support to squidlike animals, which must have been fantastically common in the Cretaceous sea that accumulated these sediments.

A close inspection of the sand and gravel yields small internal molds of several kinds of gastropods and pelecypods. Sharks' teeth and extremely rare dinosaur and pterosaur bones have been found locally but have not turned up for me except in my imagination. The illustration shows most of the kinds of fossils we found. In places, they are nearly as common as shown—but no dinosaur footprints here.

These are the fossils I collected here:

Bryozoans: unidentified encrusting species (on shells)
Cephalopods: *Belemnitella americana* (belemnoids)
Gastropods: internal molds
Pelecypods: *Anomia tellinoides; Exogyra costata; Ostrea mesenterica; Ostrea vomer; Pycnodonte (Gryphaea)* sp.; internal molds of unidentified species

Locality 42. *1*, gastropod (internal mold); *2*, pelecypod *Anomia tellinoides*; *3*, pelecypod *Ostrea vomer*; *4*, pelecypod *Ostrea* sp.; *5*, pelecypod *Exogyra costata*; *6*, pelecypod *Ostrea mesenterica*; *7*, pelecypod (internal mold); *8–10*, belemnoids *Belemnitella americana* (*10* is a base of a specimen split in two to show internal structure); *11*, pelecypod *Exogyra costata* (inner surface of shell); *12*, gastropod (internal mold); *13*, pelecypod *Ostrea* sp.; *14, 15*, pelecypods *Pycnodonte* sp. (*14* shows inner surface of shell; note that the shell wall is thick; *Anomia* (*2*) has a similar external appearance but is usually smaller and is extremely thin shelled). The dinosaur footprint is entirely imaginary. (Fossils shown .6x actual size.)

Locality 43 — along the shores of the Potomac River, Stafford and King George Counties, Virginia

Aquia formation, Tertiary period, late Paleocene epoch

Locality 43 fossils occur in beach gravel and boulders along the shore. In my experience, the best collecting is to be found for approximately 1 mile upstream and about 4 miles downstream of the mouth of Potomac Creek, which forms the border between Stafford and King George counties.

These shores may be reached by several routes. To approach the upstream area, located near Marlborough Point, take Route 608 from the town of Brooke for just over 3 miles, then turn right on Route 621. Private residences fronting on the fossiliferous shore begin to appear on the left, 1.5 miles down this road, which ends about 2.5 miles from Route 608. To get to Brooke from Stafford, turn east from Route 1 onto Route 687 (which is directly across Route 1 from the road that leads to the Stafford exchange with Interstate 95). Go 2.5 miles to a sharp right turn and continue on Route 687 for another 0.8 of a mile to the town. Route 608 is a left turn directly past the railroad overpass.

The downstream localities may be approached in two ways. Those nearest to Potomac Creek are located near Belvedere Beach, which is entirely private. To get there, take Route 3 east of Fredericksburg, turn left (northeast) onto Route 218, 0.3 of a mile east of the Rappahanock River bridge on the edge of town. Continue for approximately 6 miles to Cox Corner and turn left onto Route 600. Proceed for 4 miles, then bear left onto Route 654. Belvedere Beach is at the end of this road, a distance of 3 miles.

Fossiliferous shores may also be approached by following Route 218 for 12.4 miles east of Route 3 in Fredericksburg and turning left (north) on Route 696. After 0.9 of a mile, this road comes to the Potomac River at Fairview Beach. A campground and the Fairview Beach Crabhouse are located on the left. Access to the beach may be obtained by paying a $2.50 fee at the Crabhouse.

If you come from Route 301, turn east on Route 206, 4 miles south of the Potomac River bridge, and turn right (west) on Route 218 after 0.6 of a mile. The turn to Fairview Beach is on the right, 10.7 miles east on Route 218; the road leading to Belvedere Beach is another 6.4 miles east.

All of these localities are situated on *private property,* and *permission must be obtained before collecting.* It is a great shame that there is no guaranteed public access to this phenomenally rich and interesting collecting area. I can offer little encouragement that you will be able to get permission to collect. The Fairview Beach locality is the only access point set up for visitors. Getting to the beach elsewhere depends entirely on the goodwill of local residents.

Most of my collecting was done at Marlborough Point, which my father and I happened on by following a road until it hit water. We arrived to find a dozen people enjoying a sand and gravel beach, three or four of them looking for fossils. As soon as I reached the water's edge, I found a shark's tooth, the first of hundreds or thousands, I found over the next few years. Since that time, this access point has been closed and the road barricaded.

Although the access problem is not encouraging, it is subject to change. I have seen good waterfront collecting sites open up to visitors and close again several times. It is worth making one trip just to ask around and to try to make a friend among the landowners.

The Aquia formation was formerly identified as being early Eocene, but it is now considered to be from the late Paleocene. Its vertebrate fossils are varied and well preserved. Unlike the fossiliferous Miocene formations in the area, it has no whale bones or other marine mammal fossils. Crocodile bones and teeth are fairly common. My brother Phil found part of a jaw newly exposed in the clay with twenty-one teeth in place. This turned up at a small exposure of the bedrock about a half mile north (upstream) of Marlborough Point. I found a complete six-inch-long oyster fossil in the same stretch of beach.

The Aquia formation in this area is primarily a dark green, sandy clay, packed with white shells, most notably the high-spired gastropod *Turritella*, and a smattering of teeth and other fossils. Many boulders upstream from the Fairview Beach Crabhouse contain countless *Turritella* specimens. These boulders are quite spectacular and worth photographing for your collection.

Most of the beach gravels consist of green sandstone corkscrews, or the internal molds of *Turritella*. It is difficult to spot sharks' teeth in this coarse gravel, so the best tooth collecting is in sandier areas. On one visit, my brother Dave and I found over 300 teeth, mostly from the goblin shark *Scapanorhynchus elegans*. The most impressive sharks' teeth come from *Otodus obliquus* and attain at least two and one-half inches in slant height. I know of no place that has more sharks' teeth per square foot than here.

Fossils I have collected here are:

Birds: bone fragments
Coprolites: fossilized excrement
Corals: *Balanophyllia elaborata* (solitary species); *Haimesastraea* sp. (colonial species)
Crocodiles: *Thecachampsa* sp. (teeth and bones)
Fish (other than sharks and rays): *Paralbula* sp. (lady fish) (dental plates); *Phyllodus toliapicus* (wrasse) (dental plates); vertebrae of several unidentified species
Gastropods: *Epitonium (Scala) virginianum* (wentletrap); *Natica (Lunatia) marylandica* (moon snail); *Turritella humerosa* (turret snail); *Turritella mortoni* (turret snail)

Pelecypods: *Crassatella alaeformis* (crassatella clam); *Cucullaea gigantea* (ark clam); *Dosiniopsis lenticularis* (venus clam); *Glycymeris* sp. (bittersweet clam); *Ostrea alepidota* (oyster); *Ostrea compressirostra* (oyster); *Venericardia planicosta* (cardita clam); many more species are present
Plants: petrified wood and seeds
Rays: *Myliobatis* sp. (eagle ray) (dental plates); tail spines of unidentified species
Sharks: teeth, vertebrae, and spines; *Odontaspis* sp. (sand shark); *Otodus obliquus* or *Lamna obliqua (mackerel shark); Scapanorhynchus (Odontaspis) elegans* (goblin shark)
Turtles: shell fragments and bones; *Asperidites (Trionyx) virginiana* (soft-shelled turtle); large unidentified species

Locality 43 *(opposite)*. *1, 2,* sharks' teeth *Otodus obliquus; 3,* shark's tooth *Odontaspis* sp.; *4, 5,* sharks' teeth *Scapanorhynchus elegans; 6,* turtle shell fragment *Asperidites virginiana; 7,* coprolite; *8,* gastropod *Turritella mortoni; 9,* gastropod *Turritella* sp. (internal mold); *10, 11,* fish dental plates *Phyllodus toliapicus; 12, 13,* ray dental plates *Myliobatis* sp.; *14,* fish dental plate *Paralbula* sp.; *15,* coral *Balanophyllia elaborata; 16,* gastropod *Natica marylandica; 17,* gastropod *Epitonium virginianum; 18,* pelecypod *Crassatella alaeformis; 19–21,* crocodile teeth *Thecachampsa* sp.; *22,* crocodile (?) bone; *23,* coral *Haimesastraea* sp. (Specimens shown .6x actual size.)

Locality 43, Potomac River, Virginia

Locality 44 along the Calvert Cliffs on the western shore of the Chesapeake Bay, Calvert County, Maryland

Calvert, Choptank, and St. Mary's formations, Tertiary period, Miocene epoch

---

Locality 44, the western shore of the Chesapeake Bay extending south from Chesapeake Beach for about 25 miles to Cove Point, is littered with fossils derived from the cliffs that line much of this coast as well as from underwater outcrops. Most of this area is in private hands, and public access points are few and are restricted to the beaches south of the town of Prince Frederick. All are best approached from Maryland Route 2-4, which runs parallel to the Bay.

A good starting point is the Calvert Marine Museum in Solomons, approximately 19 miles south of Prince Frederick. This museum is located just east of the high bridge, where Route 2-4 crosses the Patuxent River. It displays many excellent fossil specimens and sells guides to local collecting. Admission is free.

1. *Calvert Cliffs State Park* is east of Route 2-4, 5.1 miles north of the museum and about 14 miles south of Prince Frederick. There is a large parking area with facilities for picnics. However, the beach is 1.8 miles away; it is reached by following the marked hiking trail. This park is open from 10 A.M. to 6 P.M. but closes on November 5 for the winter (the best season for collecting!). Admission is free.

2. *Flag Ponds Nature Park* is 4.2 miles farther north along Route 2-4, 9.3 miles north of the museum and about 10 miles south of Prince Frederick. From Route 2-4, turn east at the sign onto the gravel road. There is a charge of $3 per vehicle and a pleasant, 0.5 of a mile hike from parking to the beach. The park is open from 10 A.M. to 6 P.M. daily from Memorial Day through Labor Day and on weekends only during the rest of the year.

3. *Matoaka Cottages* is a privately owned cluster of vacation cottages, which offers year-round access to the beach for a fee: $3 a day for adults, $1 a day for

---

Locality 44a *(opposite). 1,* shark's tooth *Isurus desori; 2,* shark's tooth *Hemipristis serra; 3,* shark's tooth *Galeocerdo contortus; 4,* stone crab claw; *5, 6,* sharks' teeth *Carcharodon megalodon* (5 is missing portions of both sides of its base); *7,* shark's vertebra; *8,* crocodile tooth *Thecachampsa antiqua; 9,* pelecypod *Corbula* sp.; *10,* gastropod *Ecphora gardnerae; 11,* barnacles *Balanus concavus; 12,* drumfish jaw fragment *Pogonias* sp. (teeth are missing); *13,* gastropod *Turritella plebeia; 14,* porpoise tooth; *15,* porpoise ear bone (periotic); *16,* brachiopod *Discinisca lugubris; 17,* coral *Astrhelia palmata; 18,* pelecypod *Chesapecten nefrens; 19,* ray dental fragment *Aetobatis* sp.; *20,* ray dental plate *Myliobatis* sp. (Specimens shown .6x actual size.)

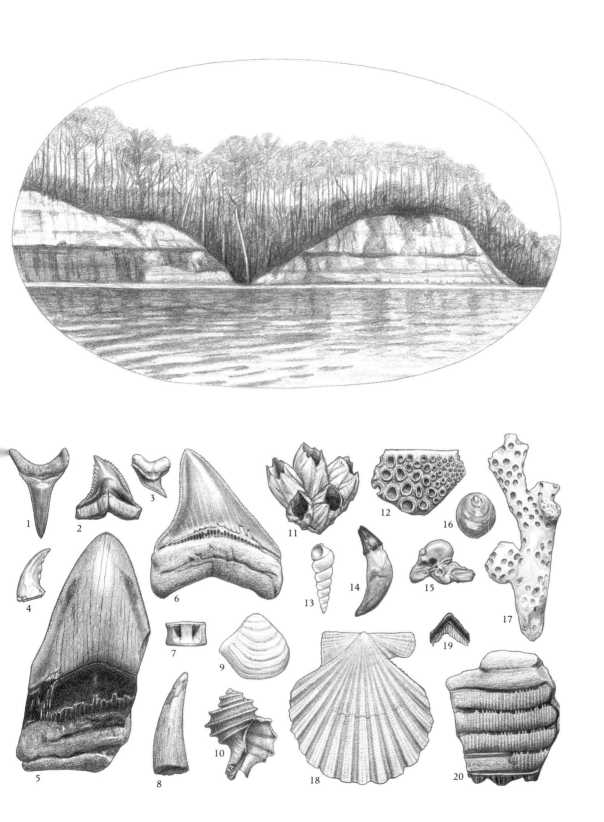

Locality 44, Calvert Cliffs, Maryland 151

children under twelve years. From Route 2-4, turn east 3 miles north of Flag Ponds (7 miles south of Prince Frederick) on Calvert Beach Road to the town of St. Leonard. Continue on Calvert Beach Road for 1.3 miles, then turn left onto a dirt road at the sign to Matoaka (pronounced MAH-toe-AAH-kah). The cottages are 0.7 of a mile down this dirt road, and the office is in front of you at its end. The owners told me that collectors who arrive before they open to catch an early tide may pay upon returning.

The Calvert Cliffs are among the best-known and most productive fossil-hunting areas in the country. Beach collecting yields an enormous variety of beautifully preserved specimens. So I have difficulty containing my indignation at the fact that there are so few points of legal acess, and only one really good one (Matoaka Cottages) that is open (for a fee) during the best season for collecting.

A whole string of communities lies perched atop these famous cliffs, each declaring in no uncertain terms that its beach is open only to residents. In my opinion, the state and local governments have let the public down by not ensuring abundant, year-round access to natural wonders that are part of everyone's heritage—like the Grand Canyon or the Smoky Mountains. The fact that the one public access point to the cliffs is nearly a two-mile hike from parking only adds salt to the wound.

Now that I've vented my spleen a bit, let me describe just how rich these hunting grounds are. Most of my collecting was done along the two-mile stretch between Governor's Run and Scientists Cliffs, two to four miles north of Matoaka Cottages. This beach used to be accessible by paying a fee but has since been closed. I have made visits to the localities described above and am convinced that the same fossils and general conditions prevail as in my old collecting grounds.

I begin with a warning: *don't dig or climb on the cliffs*! The danger can be considerable; I've heard of people being killed by falling sections of cliff, and smaller avalanches are constantly occurring. Also, it is illegal to dig into the cliffs, as they are private property.

There are three distinct formations represented by the fossils on the beaches. From oldest to youngest, these are the Calvert, Choptank, and St. Marys formations, all of Miocene age and all rich in fossils. After being eroded from the cliffs, their fossils get jumbled together as currents move and mix specimens from beach to beach. But the general trend is, the further south, the younger the fossils. Most of the specimens at the three localities mentioned above are from the Choptank and St. Marys formations.

A fabulous diversity of shells may be found here, including the large scallop *Chesapecten,* which is probably the most eye-catching fossil. Also abundant are specimens of the large barnacle *Balanus concavus.* Numerous fossils of the coral

Astrhelia palmata may be found, especially along the beach at Flag Ponds. The beautiful gastropod *Ecphora gardnerae* also turns up now and then, but the fossils that attract the most collectors are sharks' teeth. A careful search of sand and gravel deposits reveals many fine specimens and creates the impression that the Miocene waters in this area teemed with sharks. They probably did, but keep in mind that an individual shark goes through many teeth in an average life span. A study of living sharks has found that each tooth is replaced every seven or eight days. This means that one shark produces and sheds thousands of teeth a year. It is likely that prehistoric sharks went through teeth just as quickly. Whale, porpoise, and crocodile teeth are also found along the Calvert Cliffs but are comparatively rare, partly because their owners hold onto their chops a bit longer than sharks do.

Since modern pelecypods and gastropods abound in the Chesapeake Bay, their shells also litter the beaches and may be confused with fossils. Most fossil shells are chalky white and more heavily built than shells of modern local animals. Modern shells frequently bear colorful markings and are glossier than the ancient specimens. Whale bones, which are fairly common, are usually black or brown, in contrast to the white or cream colors of the bones picnickers leave behind. And there is no chance of mistaking the three- to five-inch teeth of the gi-

Locality 44b. Barnacles *Balanus concavus*. (Specimen shown 1.2x actual size.)

gantic, extinct relative of the modern man-eating great white shark *Carcharodon carcharias,* for the remains of any fish swimming today in the placid bay.

I have collected the following fossils along the Calvert Cliffs:

Barnacles: *Balanus concavus* (acorn barnacle)
Brachiopods: *Discinisca lugubris*
Bryozoans: encrusting species
Corals: *Astrhelia palmata*
Crabs: claws of stone crabs
Crocodiles: *Thecachampsa antiqua* (teeth and bones)
Echinoids: *Abertella (Scutella) aberti* (sand dollar)
Fish (other than sharks and rays): *Pogonias* sp. (drum fish) (jaw section); vertebrae of several unidentified species
Gastropods: *Ecphora gardnerae* (extinct murex snail) *Turritella plebeia* (turret snail)
Pelecypods: *Anadara staminea* (ark clam); *Chesapecten nefrens* (scallop); *Corbula* sp. (corbula clam); *Crassatella* sp. (crassatella clam); *Glycymeris* sp. (bittersweet clam); *Isognomon maxillata* (tree oyster); *Mercenaria* sp. (venus clam); *Ostrea percrassa* (oyster); many more species are present
Plants: petrified wood
Rays: *Aetobatis* sp. (duck-billed ray) (dental plates); *Myliobatis* sp. (eagle ray) (dental plates)
Sharks: teeth and vertebrae; *Alopias grandis* (thresher shark); *Carcharhinus (Carcharias) egertoni* (requiem shark); *Carcharodon megalodon* (extinct white shark); *Eugomphodus* sp. (sand shark); *Galeocerdo aduncus* (tiger shark); *Galeocerdo contortus* (tiger shark); *Hemipristis serra* (snaggletooth shark); *Isurus crassus* (mako shark); *Isurus (Oxyrhina) desori* (mako shark); *Isurus hastalis* (mako shark); *Negaprion* sp. (lemon shark); *Notorhynchus (Notidanus) primigenius* (cow shark)
Whales and porpoises: vertebrae, teeth, inner ear bones, and bone fragments of many species.

Locality 45 along the Shores of the Potomac River, Westmoreland County, Virginia

Calvert, Choptank, St. Mary's, and Eastover formations, Tertiary period, Miocene epoch

---

Locality 45 fossils may be collected on the beaches below the high cliffs along the Potomac River, about 40 miles east of the city of Fredericksburg, Virginia, especially in the vicinity of Westmoreland State Park and of Stratford Hall, the birthplace of Robert E. Lee. Both the park and Stratford Hall allow public access.

1. *Westmoreland State Park* is open year-round and admission is free except for a small parking fee (50¢ at last count) in the vacation season. It is located just north of Route 3, about 20 miles east of the town of King George, Virginia. Turn north on Route 347 at the well-marked entrance to the park and continue for about 3 miles to the fork in the road near the restaurant and park office. You can park on the beachfront by bearing left on Route 347 and continuing for 1.5 miles to its end near the marina; from here, hike downstream (east) to the cliffs on your right. Or you can park opposite the nature center by turning right at the restaurant onto Route 686. The nature center will appear almost immediately on your right. Hike through the gate beyond it to the nature trail, which is well marked and starts on the right side of the large open area. Follow this trail about 1 mile to the beach. This is a better collecting area than the first one; fossils may be found along the shore in both directions.

2. *Stratford Hall* is a restored plantation, which attracts tourists interested in the history of the Lee family. Its grounds contain many nature trails, some of which lead to the beaches downstream from Westmoreland State Park. From Route 3, continue east from the entrance to the State Park for about 1 mile and turn left (north) on Route 214, following the signs for Stratford Hall. The entrance to the plantation (Route 609) is well marked. Park at the office and ask for directions to the beach. An admission fee of $5 per adult and $2.50 for children under sixteen years is charged. Membership in the Friends of Stratford Hall is available for $20 per year. This entitles the member (and up to three companions per trip) to visit as often as desired for no additional charge. Either payment plan includes the privilege of touring the plantation. Stratford Hall is open all year.

The same formations (plus the younger Eastover beds), fossils, and general collecting conditions found at the Calvert Cliffs in Maryland hold forth here, but with greater ease of access to the beach and less competition, since this is a less-developed area. Pelecypod and gastropod shell fossils are comparatively scarcer, but vertebrate specimens are abundant and varied, including many teeth of the

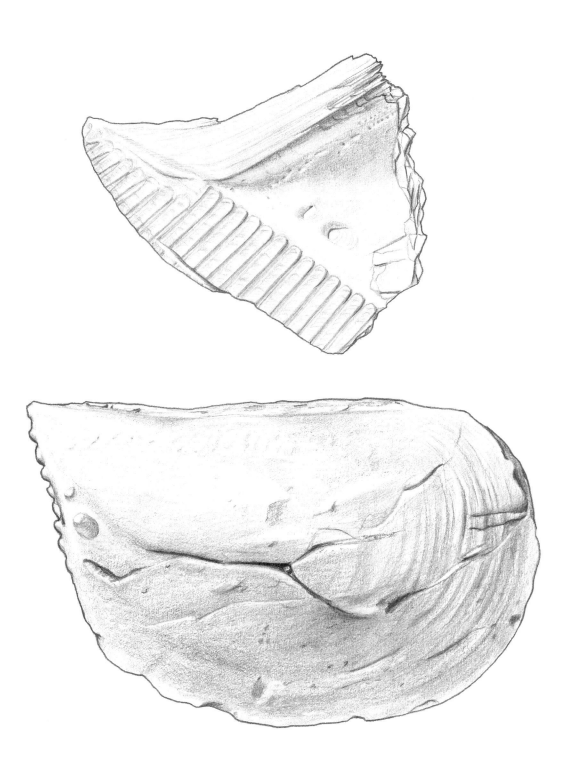

Locality 45. Pelecypods *Isognomon maxillata*. Top is a shell fragment, bottom is an internal mold, or cast. (Specimens shown .8x actual size.)

giant shark *Carcharodon megalodon.* Most of our finds were between two and four inches long, although fragments of five-inchers turned up, and we've seen some complete specimens about that size in the hands of other collectors. Examples of this kind of fossil are especially common along the Stratford Hall beach about one-half to one mile downstream of the eastern boundary of Westmoreland State Park. This same stretch of beach also produced an eighteen-inch-long rostrum, or sword, of a swordfish. It still has a sharp point and is a most impressive specimen.

Other distinctive fossils include large examples of the tree oyster *Isognomon,* which occur about one-quarter mile upstream of the eastern boundary of Westmoreland State Park. To reach these fossils, turn left at the end of the nature trail that starts near the nature center and follow the narrow beach along the cliffs *at low tide.* The water can be rough here and waist high right up to the cliffs at high tide. The clay bottom is extremely slick in places. In fact, these hazards occur all along the beach, so consult a tide chart before visiting and wear waders or hip boots if you hope to cover much distance, particularly in winter, when collecting is at its best.

Keep an eye open for flattened, oval-shaped concretions (pebble-sized or cobble-sized pieces of soft rock that wash out of the cliffs). These usually are gray, weathering to tan, and may contain beautifully preserved crab fossils, parts of which may be visible on the surface. You will have to develop a search image for these through experience; the most productive beaches are littered with vast amounts of gravel and other debris of all shapes, sizes, and colors. There are probably more shark's tooth impostors here than at any other locality I've visited.

Similar in texture and color to the concretions are coprolites, which are not common but are worth watching for, as some have fish scales inside of them. Petrified wood also turns up occasionally, and I found one horse tooth partially covered by an encrusting bryozoan. Many more invertebrate fossils occur here than are listed below; I admit to being preoccupied with the search for *Carcharodon* on most of my visits.

Stooping like a sandpiper, eyes roaming the beach for fossils that can help me see the waters of forever ago, smell the salt air, and feel the breezes that once played here.

But then my back grows weary. I straighten to gaze across the water and hear the lapping of the waves, taste the crabby evening air, watch the minnows sparkling by, the herons swooping home across the eggshell sky.

And suddenly, it's here and now—and, somehow, then as well.

I have collected the following fossils along these shores:

Barnacles: *Balanus concavus* (acorn barnacles)
Brachiopods: *Disinisca lugubris*
Bryozoans: encrusting species
Coprolites: fossilized excrement
Corals: *Astrhelia palmata*

Crabs: *Necronectes* (?) sp. (whole crabs in concretions)

Crocodiles: bony armor plates

Fish (other than sharks and rays): rostrum, or sword, of unidentified swordfish; scales and vertebrae of several unidentified species

Gastropods: *Ecphora gardnerae* (extinct murex snail); *Turritella plebeia* (turret snail); unidentified internal molds

Horse: tooth of unidentified species

Pelecypods: *Anadara* sp. (ark clam); *Astarte* sp. (Astarte clam); *Chesapecten nefrens* (scallop); *Isognomon maxillata* (tree oyster); *Lirophora* sp. (chione clam); *Mercenaria* sp. (Venus clam); many more species are present

Plants: petrified wood

Rays: spines, dermal plates (ossicles), dental plates; *Aetobatis* sp. (duck-billed ray); *Myliobatis* sp. (eagle ray)

Sharks: *Alopias grandis* (thresher shark); *Carcharhinus (Carcharias) egertoni* (requiem shark); *Carcharodon megalodon* (extinct white shark); *Eugomphodus* sp. (sand shark); *Galeocerdo aduncus* (tiger shark); *Galeocerdo contortus* (tiger shark); *Hemipristis serra* (snaggletooth shark); *Isurus crassus* (mako shark); *Isurus (Oxyrhina) desori* (mako shark); *Isurus hastalis* (mako shark); *Negaprion* sp. (lemon shark); *Notorhynchus (Notidanus) primigenius* (cow shark)

Turtles: shell fragments

Whales and porpoises: vertebrae, teeth, inner ear bones, and bone fragments of many species

Locality 46 south bank of the York River near the mouth of Indian Field Creek, York County, Virginia

Yorktown formation, Tertiary period, early Pliocene epoch

Locality 46 consists of beach deposits at the mouth of Indian Field Creek and along the south bank of the York River for at least 1 mile both upstream and downstream. The Colonial National Memorial Parkway crosses Indian Field Creek 3.7 miles west of its intersection with Route 17 in Yorktown and 12.6 miles east of its junction with Route 143 in Williamsburg.

There are large parking areas on both sides of the parkway, just east of the Indian Field Creek bridge. This bridge must be crossed on foot to reach the upstream deposits. A parking area about 1 mile to the west also provides access to the beach. Traffic may be fairly heavy on the parkway, particularly during tourist season.

The beaches of the York River in the vicinity of Yorktown are known for their abundant and varied fossils. Much of the riverfront is on U.S. Navy property and not open to collectors. Fortunately, the Colonial Parkway provides access to the river west of the town at selected points over a stretch of about seven miles.

My early collecting was done along the banks of Indian Field Creek and along the river from the parking area to the sandy point about one-quarter mile east of the bridge. The fossils in these areas are mostly worn and broken, though well-preserved sharks' teeth occasionally turn up (especially *Isurus*), and their dark-gray to black forms are easy to spot among the lighter shell fragments.

Another concentration of fossil material occurs below the cliffs along the river, about a mile east of Indian Field Creek. Shells, whales' bones, and large pieces of the coral *Septastrea* are very common here, but we have turned up only one or two sharks' teeth on our half dozen or so visits to this stretch.

The best collecting seems to be along the river upstream of the bridge. Large whale vertebrae, large sharks' teeth, enormous scallop shells *Chesapecten* (up to seven inches long), and a host of other pelecypods, gastropods (*Ecphora*), and coral fossils may be found here. There are actually two common species of *Chesapecten* to be found: *Chesapecten nefrens* and *Chesapecten jeffersonius* (which is very similar in shape but has fewer ribs).

One difficulty with this part of the beach is that collecting must be done at low tide. The shoreline, extending for a mile or so, has been covered with riprap—a jumbled mass of jagged boulders—which was dumped to retard the erosion of the river bank. At high tide, there is no sand or gravel above the waterline to walk

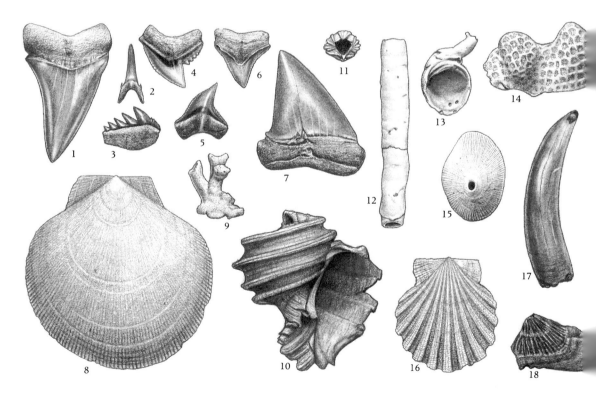

on or to scan for fossils, so consult a tide chart before visiting. It may be possible to find specimens under shallow water on calm days at high tide, but on my visits, the water has always been too active for this kind of collecting. Waders or tall boots are recommended for hunting along this stretch in cold weather, as the beach is never very wide and the tide comes in quickly. If you get trapped, you can climb over the riprap to the road above.

Despite all these problems, I can think of few beaches that are more productive. It's easy to forget your problems as you stroll along this seemingly endless shoreline and pick up the fine fossils that abound here. Sharks' teeth are not common but are unusually big for the types represented. The teeth of such species as *Galeocerdo aduncus* and *Carcharhinus egertoni* often approach or even exceed an inch in length. *Isurus* specimens frequently reach two inches, and my brother David has turned up a couple that are two and a half inches long.

By the way, Indian Field Creek lived up to its name on one of our visits. David found what at first glance appeared to be a shark's tooth but that turned out to be a black arrowhead. More artifacts from the rich human history of this area may well turn up, but the richest pickings come from a much more distant past.

These are the fossils I have collected at this locality:

Barnacles: *Balanus concavus* (acorn barnacle)
Birds: unidentified leg bones
Bryozoans: *Tretocycloecia* sp.
Corals: *Septastrea marylandica*
Crocodiles: bony plates
Fish (other than sharks and rays): *Diodon* sp. (parrot fish) (dental plate); vertebrae of several unidentified species
Gastropods: *Crepidula* sp. (slipper shell); *Diodora* sp. (limpet); *Ecphora gardnerae* (extinct murex snail); *Turritella plebeia* (turret snail)
Mammals (other than whales): triangular toe bones
Pelecypods: *Anadara* sp. (ark clam); *Chesapecten jeffersonius* (scallop); *Chesapecten nefrens* (scallop); *Crassatella* sp. (crassatella clam); *Glycymeris* sp. (bittersweet clam); *Kuphus(?)* sp. (tubes of boring clams); *Mercenaria* sp. (venus clam); *Ostrea* sp. (oyster); *Placopecten clintonius* (scallop); *Venericardia* sp. (cardita clam)
Rays: dental plates and dermal plates, or ossicles

Locality 46a *(opposite). 1,* shark's tooth *Isurus hastalis; 2,* shark's tooth *Eugomphodus* sp.; *3,* shark's tooth *Notorhynchus primigenius; 4, 5,* sharks' teeth *Galeocerdo aduncus; 6,* shark's tooth *Carcharhinus egertoni; 7,* shark's tooth *Isurus hastalis; 8,* pelecypod *Placopecten clintonius; 9,* bryozoan *Tretocycloecia* sp.; *10,* gastropod *Ecphora gardnerae; 11,* barnacle *Balanus concavus; 12,* pelecypod *Kuphus* (?) sp. (tube of boring clam); *13,* gastropod *Crepidula* sp.; *14,* coral *Septastrea marylandica; 15,* gastropod *Diodora* sp. (limpet); *16,* pelecypod *Chesapecten nefrens; 17,* sperm whale tooth *Orycterocetus crocodilinus; 18,* turtle shell fragment. (Fossils shown .6x actual size.)

Sharks: teeth and vertebrae; *Carcharhinus (Carcharias) egertoni* (requiem shark); *Eugomphodus* sp. (sand shark); *Galeocerdo aduncus* (tiger shark); *Hemipristis serra* (snaggletooth shark); *Isurus (Oxyrhina) desori* (mako shark); *Isurus hastalis* (mako shark); *Notorhynchus (Notidanus) primigenius* (cow shark)

Whales: vertebrae, teeth, inner ear bones, bone fragments of several species; *Orycterocetus crocodilinus* (sperm whale) (tooth)

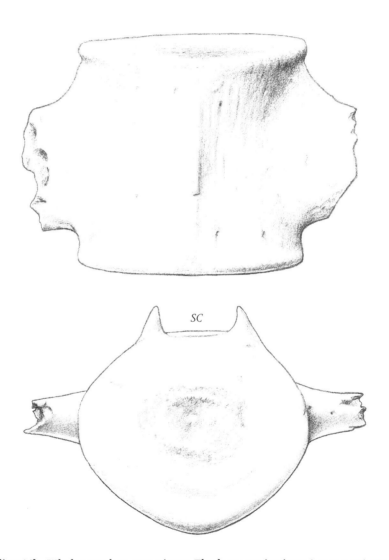

Locality 46b. Whale vertebra, two views. The large projections (processes) are mostly missing. The spinal cord (*sc*) was situated above the vertebra in the living animal, as shown. This specimen is nearly seven inches tall and weighs twelve pounds.

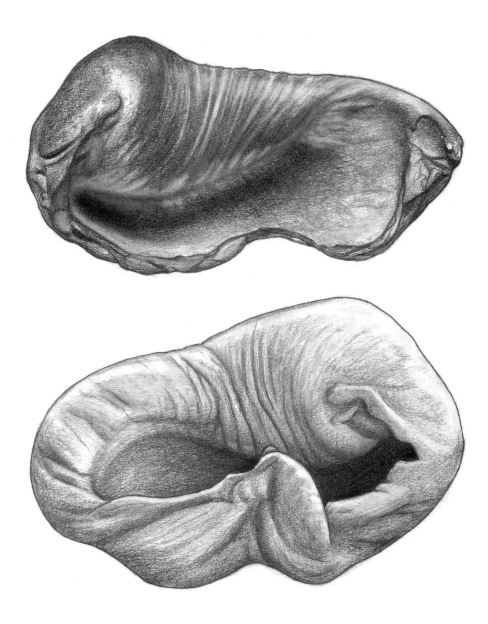

Locality 46c. Top, whale ear bone (tympanic); bottom, ear bone from a modern whale (from the opposite side of the head than the fossil). Note the parts that are missing from the fossil. (Specimens shown 1.3x actual size.)

A whale ear bone is a curious thing—dense, yet delicate. It looks like a frozen wave, as if a lifetime of listening to the swirling sea gave it its shape. There is a temptation to hold the hollow up to your ear, as you would a seashell, to hear the echo of the sounds that played here, so close to the giant mammal's brain. Do the squeals of the newborn, the joyful chatter, the moans of fear, the groans of death vibrate still in some infinite fraction in this piece of bone in my hand?

Part IV Major Fossil Groups

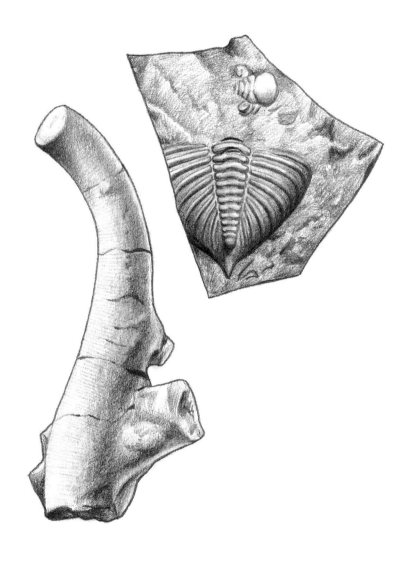

This part presents brief discussions of most of the plant and animal groups whose remains comprise the fossil record. Included are representative types that are illustrated in part 3, with locality and specimen numbers given. For example, *Amplexus* (26.5) means that specimen 5 in the illustration for locality 26 is an example of this genus. Similarly, *Astrhelia* (44a.17) refers to specimen 17 on the first illustration for locality 44, and *Odontopleura* (17b) refers to the specimen shown on the second illustration for locality 17.

Ammonites. *See under* **Cephalopods**

Ammonoids. *See under* **Cephalopods**

Anthozoans, phylum Cnidaria, class Anthozoa, Ordovician period to today. Anthozoans are members of a class of marine animals that includes corals, sea anemones, sea pens, sea fans, and sea feathers. Of these, only corals are common as fossils. Corals may be either solitary or colonial. Individual animals, called polyps, secrete chambers (called corallites in colonial forms), which often have internal radiating walls, called septa. Of the three subclasses that have left extensive fossil records—Rugosa, Tabulata, and Scleractinia—only the last group still survives.

Rugose corals, frequently called horn corals, were typically large, solitary forms, although colonial species also existed. They had well-developed septa, which were formed by the polyps in multiples of four. (Rugose corals are sometimes referred to as tetracorals, from *tetra,* meaning four.) They appeared in the Ordovician period, were abundant during the Devonian, Mississippian, and Pennsylvanian periods, and became extinct in the late Permian period.

Tabulate corals are the first corals to appear in the fossil record, occurring in early Ordovician rocks. All were colonial, and some species built extensive reefs in the Silurian and Devonian periods. Individual corallites were usually small, tubelike structures divided vertically by horizontal partitions, or tabulae. Septa were absent or poorly developed. Colonies occasionally grew to a yard or more in diameter. Tabulate corals were extinct by the end of the Permian.

Scleractinian corals appeared in the Triassic and include all living corals that secrete hard, stony skeletons. They may be solitary or colonial. Their polyps live in corallites that contain septa formed in multiples of six. (These corals are sometimes called hexacorals, from *hexa,* meaning six.) These corals built many reefs during the Mesozoic era, were less common in the early Cenozoic, and have again been building extensive reefs from the middle Cenozoic to the present day.

The following are representative anthozoan genera. (a) subclass Rugosa: *Acrocyathus* (35); *Amplexus* (26.5); *Enterolasma* (12.15,16); *Heterophrentis* (27.6,7);

Stereostylus (38.17); *Triplophyllum* (34.3,4); *Zaphrentis* (20.5). (b) subclass Tabulata: *Aulopora* (16.4); *Ceratopora* (24.3); *Favosites* (11.1,2); *Halysites* (11.3); *Pleurodictyum* (23.2), *Trachypora* (25.1). (c) subclass Scleractinia: *Astrhelia* (44a.17); *Balanophyllia* (43.15); *Septastrea* (46a.14).

Arthropods, phylum Arthropoda, Cambrian period to today. Arthropods are the enormous category of invertebrate animals that have jointed legs and external skeletons that are molted periodically. Insects are the most common living representatives of this phylum, which also includes crustaceans, spiders, millipedes, centipedes, and many others. (*See also* **Crustaceans, Eurypterids,** and **Trilobites.**)

Barnacles. *See under* **Crustaceans**

Belemnoids. *See under* **Cephalopods**

Birds. *See under* **Vertebrates**

Blastoids. *See under* **Echinoderms**

Brachiopods, phylum Brachiopoda, Cambrian period to today. Brachiopods are bivalved marine invertebrates; they were extremely abundant during the Paleozoic era, much less common in the Mesozoic, and persist in still smaller numbers and variety down to the present day. (There are about 300 living species; over 12,000 fossil species are known.) Brachiopod shells resemble those of pelecypods superficially, but these groups are not closely related. They may be distinguished from each other by the fact that the two valves, or shells, of brachiopods are almost always dissimilar in shape, while pelecypod valves are almost always mirror images of each other (oysters being the main exception).

Brachiopods are divided into two classes, which differ from each other in the way the valves are held together. The class Inarticulata is characterized by simple valves held together by muscles. The shells of members of the class Articulata are joined also through the aid of teeth and sockets, known as articulations, at the hinge. The vast majority of fossil brachiopods belong to the class Articulata.

Brachiopod shells often bear markings such as growth lines, which parallel the wide margin of the shells, and radial ribs, or plications, which radiate from the umbo, or beaklike projection at the top of each valve. Shells may also contain pits or bear spines, some of which may be very long in fossil specimens. The shells of brachiopod fossils are frequently still joined together. This is largely because the living animals must exert muscular effort to open their valves, which shut tightly when the animals die.

Living brachiopods may be free moving, or lie unattached on the sea bottom, or be attached to objects by a stalklike appendage (pedicle). The hole through which the pedicle extends may be seen on the inner surface of the umbo of one of the shells. This shell is called the pedicle valve. Several orders of articulate brachiopods are recognized, distinguished from each other mainly by differences in the shape of the structure inside the shell (brachidium) to which the soft filter-feeding organs (lophophores) were attached. The shell that bears the brachidium is known as the brachial valve.

The following are representative brachiopod orders and genera. (a) class Inarticulata—order Acrotretida: *Discinisca* (44a.16); order Lingulida: *Lingula* (15a.5); (b) class Articulata—order Orthida: *Onniella* (6.9,10); *Rhipidomella* (16.11,12); *Tropidoleptus* (24.5); order Rhynchonellida: *Cupularostrum* (8.9); *Eatonia* (15a.6,7); *Orthorhynchula* (5.9,10); order Spiriferida: *Ambocoelia* (27.14); *Athyris* (27.9); *Coelospira* (20.14); *Composita* (36.9,10); *Cyrtina* (16.20); *Delthyris* (26.3,4); *Desquamatia* (20.15); *Meristella* (16.10); *Mucrospirifer* (23.3); *Spinatrypa* (26.2); *Spinocyrtia* (24.1,2); *Spirifer* (36.11,12); order Strophomenida: *Chonetinella* (38.3); *Devonochonetes* (24.7); *Diaphragmus* (36.13,14); *Inflatia* (34.1,2); *Leptaena* (12.7); *Linoproductus* (36.8); *Orthotetes* (36.15); *Schuchertella* (20.11); order Terebratulida: *Rensselaeria* (15b.13,14).

Bryozoans, phylum Bryozoa, Cambrian period to today. Bryozoans are tiny, colonial invertebrates who live in both salt and fresh water. Bryozoan colonies may resemble those of corals, but they are more closely related to brachiopods. The individual chamber inhabited by a single animal—called a zooecium (plural, zooecia)—is generally much smaller than a corallite in a coral colony, seldom exceeding one millimeter in diameter. Like corals, individual animals are capable of gathering food by extending tentacles, which may be withdrawn when danger threatens. Fossil bryozoans often encrust the surfaces of other fossils, though free-living forms are also common, including twiglike, branching colonies and massive, often globular colonies. Thin, lacelike forms were very common in the mid to late Paleozoic.

Bryozoans are divided into three classes, one of which lives only in fresh water. The six orders of the other two classes have many representatives in the fossil record, but identification of specimens to genus or species level is difficult because classification usually depends on minute details of internal structure. Representative bryzoan genera are *Fenestella* (36.3,4); *Prasopora* (6.1–5); *Ptilodictya* (24.4); *Rhinidictya* (6.13); *Tretocycloecia* (46a.9).

Cephalopods, phylum Mollusca, class Cephalopoda, Cambrian period to today. Cephalopods are the group of marine mollusks that includes living forms such as squids, octopuses, and the chambered nautilus, as well as extinct forms such as ammonoids and belemnoids. Living cephalopods, unlike most other mollusks, have excellent vision, are capable of rapid, coordinated movement, and display well-developed intelligence. It is probable that many fossil forms also had these characteristics. Nearly all fossil cephalopods belong to three main groups: nautiloids, ammonoids, and belemnoids.

Nautiloids have chambered shells, the chambers being divided from each other by simple, disc-shaped walls called septa, and connected to each other by a tube known as a siphuncle. One Ordovician species with a straight, cone-shaped shell attained lengths of twelve feet. The chambered nautilus is the only surviving member of this group, which first appeared in the Cambrian period and was most abundant during the Ordovician period.

Ammonoids differ from nautiloids in the greater complexity of the margins of their septa, visible on the surface of the shell where the chambers meet at wavy lines called sutures. (These sutures are straight or only slightly curved in nautiloids.) Ammonoids also differ from nautiloids in the position of their siphuncles. They appeared in the Devonian period and were especially common in the Mesozoic era, becoming extinct at the end of the Cretaceous period. Different groups of ammonoids are recognized on the basis of differences in the shapes of the sutures, including the goniatites, ceratites, and ammonites.

Belemnoids were closely related to the modern squids. Their bullet-shaped internal shells (or guards) gave stability to their streamlined bodies, which were adapted for rapid swimming. Occasionally, less durable parts of the internal structure are preserved, which show chambers similar to those in nautiloid shells—but these were inside the animal; rather than vice versa. Belemnoids appeared in the Mississippian period, were most abundant in the Mesozoic, and finally disappeared in the Eocene epoch.

Representative cephalopod subclasses and genera are (a) subclass Nautiloidea: *Michelinoceras* (8.10,11; 17a.13,14); (b) subclass Ammonoidea: *Agoniatites* (18.7,8); (c) subclass Coleoidea; order Belemnoidea: *Belemnitella* (42.8–10).

Clams. *See under* **Pelecypods**

Conodonts, toothlike remains of mysterious, probably wormlike animals. These fossils are extremely small and widespread in marine sediments from the Paleozoic era. They are useful in identifying and correlating rock formations.

Conularids, phylum Cnidaria(?), order Conulariida, Cambrian to Triassic periods. Conularids are extinct marine animals that may have been related to jellyfish. Their fossils are cone- or pyramid-shaped, with very fine transverse ridges on their surfaces. A representative genus is *Conularia* (18.1,2).

Coprolites, fossilized excrement, often containing recognizable traces of other organisms within them, including fish scales, bones, hair, and feathers.

Corals. *See under* **Anthozoans**

Crabs. *See under* **Crustaceans**

Crinoids. *See under* **Echinoderms**

Crocodiles. *See under* **Vertebrates**

Crustaceans, phylum Arthropoda, superclass Crustacea, Cambrian period to today. Crustaceans include crabs, lobsters, barnacles, shrimps, ostracodes, and many other arthropods. Some members of this large group live in fresh water and some are terrestrial, but most fossils are of marine species, especially of crabs, ostracodes, and barnacles.

Ostracodes are tiny, bivalved animals whose fossils are extremely abundant in some rocks. The minute, shrimplike crustacean lives within its protective shells, which are somewhat similar to pelecypod valves in general shape. Over 10,000 fossil species have been recognized; about 200 living species are known. Most forms are marine, but freshwater—and even terrestrial—species exist today. Ostracode fossils are often useful for correlating rock layers. Though some fossil forms are as much as one inch in length, the vast majority of specimens are around one millimeter (one-fiftieth of an inch) long and are best appreciated when viewed under magnification.

Barnacles, which belong to the class Cirripedia, first appeared in the Silurian period, or earlier, and are common today. The animals themselves are shrimplike and rarely preserved, but their calcareous shells may be common as fossils.

Crabs belong to the class Malacostraca, members of which appeared in the Cambrian period, although crabs themselves did not evolve until the Jurassic period. Parts of claws are fairly common as fossils, and occasionally whole animals turn up in concretions.

Representative crustacean classes and genera are (a) class Ostracoda: *Kloedenella* (8.5,6); *Leperditia* (10.4); *Zygobolbina* (9.1,8); (b) class Cirripedia: *Balanus* (44a.11); (c) class Malacostraca: stone crab claw fragment (44a.4).

Cystoids. *See under* **Echinoderms**

Echinoderms, phylum Echinodermata, Cambrian period to today. Echinoderms are strictly marine invertebrates; the group includes sand dollars, sea urchins, other echinoids, crinoids, cystoids, and blastoids, as well as edrioasteroids, starfish, brittle stars, and sea cucumbers. There are approximately 6,000 species living today.

Echinoids first appeared in the Ordovician period and are common today. They include the sea urchins, sand dollars, and heart urchins. As in most echinoderms, a five-part radial symmetry is evident on the surface of the shell, which is made up of many plates. Some echinoids bear spines on their surfaces, which are commonly found as fossils separate from the main body. Echinoid fossils are fairly rare in Paleozoic rocks but may be common in Mesozoic and Cenozoic strata.

Crinoids first appeared in the Ordovician period and are extremely abundant as fossils in many Paleozoic formations. Most Paleozoic crinoids were attached to the sea bottom by a segmented stalk made up of usually circular plates, or segments. At the top of the stalk was a crown, or calyx, with branching arms (brachioles), which bore many small armlets (pinnules) used in gathering food. Stalked crinoids, which are often called sea lilies, still survive today, but stalkless forms, called feather stars, are much more common in modern seas.

Cystoids are also stalked like crinoids (with rare exceptions). Their crowns tend to be somewhat pear-shaped and are perforated by small openings. Cystoids appeared in the Ordovician period and were gone by the end of the Devonian period.

Blastoids are a third group of stalked echinoderms. Their symmetrical crowns bear distinctive fivefold markings on their tops; from which delicate arms extended in life, although these structures are rarely preserved. Blastoids lived from the Silurian to Permian periods.

Edrioasteroids were small, disc-shaped animals that resembled tiny starfish sitting in saucers. Their fossils are occasionally found in rocks of Cambrian to Mississippian age.

Starfish and brittle stars (class Stelleroidea) both appeared in the Ordovician period and still survive today. Fossils usually consist of disarticulated plates and pieces, neither of which are common in the Mid-Atlantic region. Sea cucumbers (class Holothuroidea) are also prone to breaking into small plates after death. Also, much of their bodies consists of soft tissues, so these are not common as fossils.

Representative Echinoderm classes and genera are (a) class Crinoidea: *Arthroacantha* (21.4–7; 25.2); *Edriocrinus* (16.18,19); *Ectenocrinus* (3.11); (b) class Cystoidea: *Pseudocrinites* (10.1); (c) class Blastoidea: *Pentremites* (34.14).

Eurypterids, phylum Arthropoda, class Merostomata, subclass Eurypterida, Cambrian to Permian periods. These extinct marine animals are also known as sea scorpions. They attained gigantic size, especially in the Silurian period, when individuals grew to nine feet. The majority of fossil specimens are one foot or less in length, however. Not common as fossils, the best specimens in the United States probably come from New York State, where one species has been dubbed the official state fossil.

Fish. *See under* **Vertebrates**

Gastropods, phylum Mollusca, class Gastropoda, Cambrian period to today. This large group includes snails and slugs, although only the former is important to fossil collecting. Snail shells may superficially resemble cephalopod (ammonoid or nautiloid) shells but are not chambered. Also, snail shell spirals generally form a pyramidal or conical form, while cephalopod shells almost always coil in a plane, producing a flat shell. Gastropods, which may be marine, freshwater, or terrestrial, first appeared in the Cambrian period and are probably as abundant today as they ever were in the past. (*See also* **Pteropods**). Representative gastropod genera are: *Bembexia* (27.15); *Crepidula* (46a.13); *Ecphora* (46a.10); *Epitonium* (43.17); *Loxonema* (27.11,12); *Murchisonia* (33.1); *Natica* (43.16); *Platyceras* (15a.2–5); *Shansiella* (38.15,16), *Turritella* (43.8,9).

Graptolites, phylum Hemichordata, class Graptolithina, Cambrian to Mississippian periods. Graptolites were colonial marine animals. Their fossils usually consist of flat, dark, linear films on rock surfaces, especially in shales. They are particularly common in Ordovician strata. Many graptolites were free floating, and their fossils often occur in rocks far from areas where other marine organisms lived. Because of this, they have great value for the correlation of lower Paleozoic rocks. A representative graptolite genus is: *Climacograptus* (5.1).

Hyolithids, phylum unknown, order Hyolithida, Cambrian to Permian periods. Hyolithids are small to medium-sized, cone-shaped fossils found in Paleozoic rocks. Their relationship to other organisms and their manner of living are unknown, but some paleontologists believe they were related to the mollusks. Hyolithids were abundant during the Cambrian period, but uncommon in later Paleozoic rocks. A representative genus is: *Hyolithes* (20.6).

Nautiloids. *See under* **Cephalopods**

Ostracodes. *See under* **Crustaceans**

Pelecypods, phylum Mollusca, class Pelecypoda, Cambrian period to today. Pelecypods, which are assigned to the class Bivalvia by some scientists, are bivalved mollusks. They are represented today by a wide variety of forms, including clams, mussels, oysters, scallops, and many more. Most pelecypod fossils consist of individual shells. This is partly due to the fact that the animal uses muscles to hold its shells together in life. When the pelecypod dies, its shells open naturally and are easily dispersed.

Pelecypod shells often bear a complicated assortment of teeth and sockets along the hinge line. These keep the valves from rotating relative to each other. Many shells from Tertiary deposits have single holes on their surfaces. These usually mean that the pelecypod was killed by a predatory snail, which bored through the shell to get to the soft animal within. The internal surface of pelecypod valves also bear rounded marks where muscles were attached in life; these are called muscle scars. The arrangement of muscle scars, as well as of the teeth and sockets, is important in the classification of pelecypods. Most pelecypods are marine, though many freshwater species also exist.

Representative pelecypod genera are: *Ambonychia* (5.5,6); *Anomia* (42.2); *Chesapecten* (44a.18); *Cleidophorus* (8.4); *Corbula* (44a.9); *Cornellites* (23.1); *Crassatella* (43.18); *Cypricardella* (24.6); *Exogyra* (42.5,11), *Nuculopsis* (38.4,5); *Orthonota* (27.16); *Ostrea* (42.3,4,6); *Phestia* (38.14), *Praecardium* (17a.10); *Ptychopteria* (24.8); *Pycnodonte* (42.14,15), *Wilkingia* (36.1).

Plants, kingdom Plantae, Precambrian eras to today. Plants have had a long and rich evolutionary history, but the fossil record is slanted in favor of those forms that lived near water, where sediments accumulated. Recognizable fossils of fully terrestrial plants are scarce in most places and consist primarily of microscopic pollen grains and "petrified" trunks of trees, which may or may not be identifiable.

Most plant fossils common in the Mid-Atlantic region are of the species that inhabited the coal forests of the Pennsylvanian period. These include horsetails (class Sphenopsida), scale trees (class Lycopsida), true ferns (class Phyllicopsida), seed ferns (class Gymnospermopsida, order Pteridospermales), and the probable ancestors of conifers belonging to the genus *Cordaites* (class Gymnospermopsida, order Cordaitales). All of these classes belong to the phylum Tracheophyta, which also includes angiosperms, or the flowering plants (class Angiospermopsida). Angiosperms appeared in the Cretaceous period and include the majority of living species. Members of this group are as diverse as oaks, grasses, and roses.

Horsetails, or sphenopsids, first appeared in the Devonian period and attained heights of nearly 100 feet in the Pennsylvanian period. Smaller species still survive today. Most horsetail fossils consist of leaves and internal molds of stems. Scale trees (lycopsids) were the giants of the Paleozoic coal forests, growing to as much as 150 feet. They first appear in Devonian rocks and are especially common in Pennsylvanian rocks. Tiny relatives still survive. The leaves of seed ferns (pteridosperms) resemble those of true ferns in shape and arrangement but they bear seeds rather than spores. Seed ferns lived from the Mississippian to the Jurassic periods. Cordaitales evolved by the Mississippian period and became extinct in the Permian period. They had long, straplike leaves and are probably ancestral to the modern conifers (class Gymnospermopsida), which include pines, cedars, redwoods, and spruces—the cone-bearing plants.

Fossils of Mesozoic plants such as sphenopsids, cycads (class Gymnospermopsida), and ginkgoes (class Gymnospermopsida) have been found in the Triassic and Cretaceous rocks of the Mid-Atlantic area but not at the localities in this book. Cycads are palmlike plants; they were very common in the Mesozoic era and survive in tropical areas today. Only one species of ginkgo still survives; its leaves are virtually indistinguishable from those of some Mesozoic species.

Some marine algae have the ability to secrete calcium carbonate. The structures that these plants produce are readily preserved as fossils. Two very different kinds of algae have this capability. Primitive blue-green algae (phylum Cyanophyta) commonly built pillowlike structures called stromatolites in the late Precambrian era and into the Cambrian period. These structures are exceedingly rare thereafter, though stromatolites are still forming today on the west coast of Australia.

Some species of the more advanced algae known as green algae (phylum Chlorophyta) also can secrete calcium carbonate. Fossils of these plants occur as globular masses or encrusting sheets, frequently attached to the surface of other fossils. Another type of green algae is believed to have formed strange fossils known as receptaculitids. Specimens consist of a spiralling pattern of plates and pillars or pits over a bowl-shaped surface. These fossils are restricted to Paleozoic rocks.

Plant classification is a variable science. Many scientists divide the plant kingdom into divisions rather than phyla. Another oddity in the naming of plants is that different names are sometimes used for different parts of plants of the same genus. Thus, the leaves of Pennsylvanian horsetails are called *Annularia*, while stems are referred to the genus *Calamites*. Similarly, the roots of the scale tree *Lepidodendron* are called *Stigmaria*, its leaves *Lepidophyllum,* its conelike reproductive structures *Lepidostrobus*, even though fossils show that all are parts of the same plant. Representative classes and genera of plants in the phylum Tracheophyta are (a) class Sphenopsida: *Annularia* (41.5); *Calamites* (37.5); (b) class Lycopsida:

Lepidodendron (37.9,11); *Lepidophyllum* (40.5); *Lepidostrobus* (37.1); *Stigmaria* (37.2); (c) class Phyllicopsida: *Alloiopteris* (40.3); *Pecopteris* (40.1,8); (d) class Gymnospermopsida, order Pteridospermales: *Alethopteris* (37.4); *Mariopteris* (37.10); *Neuropteris* (41.1–4,6,7); *Trigonocarpus* (37.6,7).

Poriferans, phylum Porifera, Precambrian era to today. More commonly known as sponges, poriferans are seldom well preserved as fossils. In some rocks, their tiny, siliceous hard parts, known as spicules, may be fairly common. Their presence may also be inferred from the numerous holes they have bored in gastropod and pelecypod shells. A representative poriferan specimen is the unidentified species at (12.17).

Pteropods, phylum Mollusca, class Gastropoda, order Pteropoda, Cambrian(?) period to today. Modern pteropods, also known as sea butterflies, are planktonic snails found in the open ocean. Some problematic, needle-shaped Paleozoic fossils have been tentatively ascribed to this group.

Rays. *See under* **Vertebrates**

Receptaculitids. *See under* **Plants**

Rugose corals. *See under* **Anthozoans**

Sharks. *See under* **Vertebrates**

Snails. *See under* **Gastropods**

Sponges. *See under* **Poriferans**

Stromatolites *See under* **Plants**

Stromatoporoids, phylum Porifera?, class Stromatoporoidea, Cambrian period to Cretaceous period. These extinct marine organisms built layered, often massive structures that formed reefs, especially during the Silurian and Devonian periods. The classification of stromatoporoids with poriferans is very tentative, since so little is known about them. Two representative stromatoporoid specimens are shown at: (12.21; 16.1).

Tentaculitids, phylum unknown, order Tentaculitida, Ordovician period to Devonian period. The tentaculitids are another problematical group of Paleozoic marine invertebrates. They occur in great numbers in some rocks, often stacked on top of each other. Their small, elongated cones often have distinctive rings along their lengths. These fossils may superficially resemble straight-shelled nautiloids but are not chambered and are usually much smaller. A representative genus is *Tentaculites* (15b.11).

Trace Fossils, Precambrian era to recent. Trace fossils are the marks left by ancient organisms rather than their physical remains. These include burrows, trails, coprolites (fossilized excrement), nest holes, even the scour marks made by plants rubbing across mud that are sometimes preserved in shale. All are indirect evidence of past life.

Trilobites, phylum Arthropoda, class Trilobita, Cambrian period to Permian period. There are approximately 4,000 known species of trilobites. These marine invertebrates were abundant in the early Paleozoic era and dwindled in numbers and variety as that era progressed. Of the nine orders of trilobites, only one survived past the Devonian.

The three lobes referred to by the name tri-lobite are counted from side to side and are not the head (or cephalon), thorax, and tail (or pygidium). Several species exceeded one foot in length, and at least two are known to have exceeded two feet; but most adult trilobites were between one and three inches long.

The majority of trilobite fossils represent the discarded parts of the exoskeleton left behind after molting. Complete specimens are rare, and remains of the delicate limbs and antennae much rarer still. One of the closest living relatives of the trilobites is believed to be the horseshoe crab (*Limulus*). Its habits suggest the possibility that at least some trilobites swam upside down, as this crab often does.

Representative trilobite genera are: *Ampyxina* (2.1,3); *Basidechenella* (18.3); *Calymene* (9.2–5,11); *Coronura* (20.17,18); *Cryptolithus* (4); *Dalmanites* (15a.1,10), *Dionide* (2.2), *Flexicalymene* (3.1–3); *Greenops* (27.1); *Odontocephalus* (17a.1,2); *Odontopleura* (17b); *Phacops* (20.8–10); *Trimerus* (15b.15; 27.3).

Turtles. *See under* **Vertebrates**

Vertebrates, phylum Chordata, subphylum Vertebrata, Cambrian period to today. Though vertebrates have been abundant since the Ordovician period, their fossils are not common. This is due partly to the fact that many kinds were terrestrial and died in places that were not conducive to the preservation of their remains and partly to the fact that the bones of their skeletons are held together by soft tissues, which disintegrate rapidly after death. Because of this latter factor, vertebrate bones are often more easily scattered and destroyed than invertebrate shells might be.

Most vertebrate fossils are the teeth of marine forms, and none are more common than the teeth of sharks and rays (class Chondrichthyes). Sharks first appeared in the Devonian period and, of course, continue to flourish. The best-known fossil shark is the extinct relative of the great white, or maneater, shark, *Carcharodon*. Teeth from the fossil species attain eight inches in length. The living animals are estimated to have approached fifty feet.

Rays (and their relatives, the skates) first lived in the Pennsylvanian period. Their teeth, which are often fused into flattened plates for crushing shellfish, are especially abundant in Tertiary sediments.

Fossils of the so-called bony fish (class Osteichthyes) are not common in the Mid-Atlantic region, though isolated teeth and vertebrae turn up fairly often in Tertiary formations. They are referred to as bony because of the greater hardness, or ossification, of their skeletons than is found among the Chondrichthyes, whose skeletons are composed primarily of flexible cartilage. Complete specimens of bony fish fossils have been found in the Triassic rocks of Virginia and in some Paleozoic formations, but are extremely rare finds, and the locations of productive sites are jealously guarded secrets. Fish from this group first appeared in the early Devonian period and comprise the majority of living species, including such familiar types as bass, tuna, sailfish, and sardines.

Amphibian fossils (class Amphibia) have been found in some of the late Paleozoic rocks of the Mid-Atlantic area, but are even rarer than fish. They first appeared in the Devonian period and grew to large sizes in the coal forests of the Pennsylvanian and early Permian periods.

The best known prehistoric reptiles (class Reptilia) are, of course, the dinosaurs. But their fossils are hard to come by in most areas, including the Mid-Atlantic region. Fossils of other reptiles are reasonably common in Tertiary deposits, particularly the bones and teeth of crocodiles (order Crocodilia) and fragments of turtle shells (order Chelonia). Reptiles first appear in Pennsylvanian rocks.

Bird bones (class Aves) are characteristically hollow with thin walls. This reduces body weight and aids in flight. However, it also makes the bones fragile and less likely to be fossilized. Nevertheless, specimens may be found in Tertiary formations in the Mid-Atlantic area. They can usually be distinguished from modern remains by their greater density (due to mineralization), greater brittleness, and especially by their darker, more uniform color. (These criteria may also be applied to other vertebrate remains.) Birds first appeared in the Jurassic period and have been abundant since the Cretaceous period.

Remains of mammals (class Mammalia) are very scarce in the Mid-Atlantic area, with the exception of marine mammals of the order Cetacea (whales, dolphins, and porpoises). Whale bones are fairly common in late Tertiary formations of the Coastal plain, as are fossils from their smaller relatives. Some teeth and the occasional complete skeleton have also turned up. Terrestrial mammal fossils, such as horse or mastodon teeth and bones, occur very rarely in Tertiary marine deposits. Presumably, they are from carcasses that were washed into the sea. Similar fossils are found in scattered places, such as caves. Mammals first appeared in the late Triassic period but did not become abundant and varied until the Tertiary period.

The following are representative vertebrate classes, orders, and genera. (a) Class Chondrichthyes—order Selachii (sharks): *Carcharhinus* (46a.6); *Carcharodon* (44a.5,6); *Eugomphodus* (46a.2); *Galeocerdo* (46a.4,5); *Hemipristis* (44a.2); *Isurus* (46a.1,7); *Notorhynchus* (46a.3); *Odontaspis* (43.3), *Otodus* (43.1,2), *Scapanorhynchus* (43.4,5); order Rajiformes (rays): *Aetobatis* (44a.19); *Myliobatis* (43.12,13). (b) Class Osteichthyes: *Paralbula* (43.14); *Phyllodus* (43.10,11); *Pogonias* (44a.12). (c) Class Amphibia—subclass Labyrinthodontia: jaw fragment (38.10). (d) Class Reptilia—order Crocodilia: *Thecachampsa* (43.19–21; 44a.8); order Chelonia: Asperidites (43.6). (e) Class Mammalia—order Cetacea: *Orycterocetus* (46a.17); representative vertebra (46b).

Appendix

Annotated Bibliography

Index

Appendix:
Geological Surveys in the Mid-Atlantic Region

National Cartographic Information Center
U.S. Geological Survey
Mail Stop 507
National Center
Reston, VA 22092

Delaware Geological Survey
University of Delaware
Newark, DE 19711

Maryland Geological Survey
Merryman Hall
Johns Hopkins University
Baltimore, MD 21218

Pennsylvania Bureau of Topographic and Geologic Survey
Department of Environmental Resources
Box 2357
Harrisburg, PA 17120

Virginia Division of Mineral Resources
Box 3667
Charlottesville, VA 22903

West Virginia Geological and Economic Survey
Box 879
Morgantown, WV 26507

Annotated Bibliography

For many years, there was a shortage of popular field guides to fossil identification, but recent volumes have done much to rectify the problem. Specifically, useful volumes have been published by Macmillan, Simon and Schuster, and the National Audubon Society with Alfred A. Knopf. However, none of these does an adequate job on common vertebrate fossils. Some of the field guides to collecting published by individual states are very good; the one put out by the State of Pennsylvania is excellent. These publications often list localities for collecting.

If I had to single out one book for overall brilliance, it would be the *Audubon Society Field Guide to North American Fossils*, by Ida Thompson. It contains a wealth of information and illustrations and lacks only descriptions of specific localities and in-depth coverage of vertebrate forms. Other particularly useful books are *Index Fossils of North America*, by Shimer and Shrock, a reference book with illustrations of many hundreds of invertebrate fossils, and *Fossils for Amateurs*, by MacFall and Wollin, which presents many techniques for collecting and preparing specimens.

In addition to valuable scientific information, paleontology and historical geology textbooks often contain many pictures of fossils and reconstructions of past life. Current editions may be quite expensive, but out-of-date volumes can often be had for a song at used book stores. Guidebooks to shells and living invertebrates can provide much information about the organisms that left fossils behind, especially those from Tertiary deposits, as many of the same genera still live in the region today. In some cases, these books can even be helpful for identification. Helpful information can also be found in the catalogs issued by organizations that make a business of selling fossils. These often contain fairly extensive lists of specimens, giving the geologic ages, rock formations, and general localities where they were found—details that may help solve a sticky identification problem.

In addition to the books listed below, there are numerous professional papers and journals available. Much depends on being able to identify your specimens, so the more source material you have, the easier and more enjoyable that will be. (Note: I have put an asterisk before the books I consider most valuable to the collector.)

Popular Guides to Fossils and Fossil Collecting

*The Audubon Society Field Guide to North American Fossils, by Ida Thompson (New York: Alfred A. Knopf, 1982). Probably the best one book to have; it presents many useful facts and full-color photographs of specimens in a compact package. An excellent guide to the hobby.

The Dawn of Life, by Giovanni Pinna (New York: World Publishing, 1972). An atlas of full-color photographs of fairly sensational fossils from around the world.

Eyewitness Books—Dinosaur, by D. Norman and A. Milner (New York: Alfred A. Knopf, 1989). A brilliantly illustrated book, mentioned here because of its emphasis on actual dinosaur fossil specimens, shown in clear, color photographs.

Familiar Fossils (An Audubon Society Pocket Guide), by Sidney Horenstein (New York: Alfred A. Knopf, 1988). A small but well-chosen and beautifully photographed selection of North American fossils. The descriptions include interesting facts about the animals' lifestyles.

Fossils, by Mark Lambert (New York: Arco Publishing, 1978). Doesn't show enough fossils to serve as an identification guide but is full of color reconstructions of ancient life and photographs of some impressive fossils.

**Fossils,* by Richard Moody (New York: Macmillan, 1986). An attractive guide to the common fossils of North America and Europe. Contains color photographs of over 300 genera. An unusual feature is a period-by-period discussion of prehistoric life, with sample fossils for each.

Fossils: A Guide to Prehistoric Life, by F. Rhodes, H. Zim, P. Shaffer (New York: Golden Press, 1962). An old standby and a good place to start but not comprehensive enough to take you very far.

*The Fossil Book (rev. ed.), by C. Fenton, M. Fenton, P. Rich, and T. Rich (New York: Doubleday, 1989). A beautifully conceived, written, and illustrated book, and an excellent introduction to the hobby. Originally published in 1958, it has been updated and expanded—and given an expensive price.

**Fossils for Amateurs* (rev. ed.), by Russell P. MacFall and Jay C. Wollin (New York: Van Nostrand Rheinhold, 1983). A wonderful handbook for the hobby, full of techniques and basic information.

Fossils in Colour, by J. Kirkaldy (London: Blandford Press, 1967). A field guide to fossil hunting in Great Britain, which is widely available in the States. Worth having for its informative and well-written text. A small but well-photographed collection of fossils is shown, including the ugliest *Carcharodon* ever found.

Fossils of the World, by V. Turek, J. Marek, J. Benes (New York: Arch Cape Press, 1989). A beautiful volume with over 800 color photographs of excellent specimens. However, the title is misleading, as most of the fossils are from Europe—the majority from Czechoslovakia. But nice to browse through.

**Fossil Sharks: A Pictorial Review,* by Gerard R. Case (Jersey City, N.J.: Gerard R. Case, 1973). An excellent guide to fossil sharks' teeth, with many good quality photographs. A nice touch is the inclusion of examples of teeth from living sharks and rays.

**Fossil Shark and Fish Remains of North America,* by Gerard R. Case (New York: Gerard R. Case, 1967). A very well illustrated publication, showing a broad range of these kinds of fossils.

**Fossil Vertebrates—Beach and Bank Collecting for Amateurs,* by M. C. Thomas (Venice, Fla.: M. C. Thomas, 1968). A very handy book for identifying unusual vertebrate fossils you'll find on the beach.

Handbook of Fossil Collecting, by Gerard R. Case (Jersey City, N.J.: Gerard R. Case, 1972). A slim, nicely illustrated introduction to fossil collecting, with a few localities.

The Henry R. Holt Guide to Minerals, Rocks, and Fossils, by W. Hamilton, A. Woolley, and A. Bishop (New York: Henry R. Holt, 1989). The last third of this book presents color photographs and descriptions of many fine fossils. Some specimens are shown a bit too small for details to be seen, however.

The Observer's Book of Fossils, by Rhona M. Black (London: Frederick Warne, 1977). A British publication, but worth having because most of the fossils occur in the United

States as well. It contains excellent discussions of the probable life modes of the animals.

A Pictorial Guide to Fossils, by Gerard R. Case (New York: Van Nostrand Rheinhold, 1982). A pricey but unparalleled collection of 1,300 illustrations of all kinds of fossils. Especially good for vertebrate material.

Simon and Schuster's Guide to Fossils, by P. Arduini and G. Teruzzi (New York: Simon and Schuster, 1986). A beautifully photographed collection of unusually nice fossils. This book contributes useful information about the distribution of fossil types and does a better job on vertebrates than most guides.

Trilobites: A Photographic Atlas, by Riccardo Levi-Setti (Chicago: University of Chicago Press, 1975). Not very useful for identification as it lacks an index, but magnificently photographed. Sit back, relax, and marvel at the beauty and fascination of trilobites.

Trilobites of the Thomas T. Johnson Collection—How to Find, Prepare, and Photograph Trilobites, by Thomas T. Johnson (Morrow, Oh.: Thomas T. Johnson, 1985). An exuberant, very well illustrated presentation of a collection that will leave you breathless. Available through the Smithsonian Institution, Museum of Natural History Bookstore.

Guides to Collecting Localities

Appalachian Mineral and Gem Trails, by Jane Culp Zeitner (San Diego: Lapidary Journal, 1968). This book is aimed at the gem and mineral collector but lists some fossil localities as well. I have accumulated some nice mineral specimens over the years while fossil collecting; this hobby can become an interesting sideline.

Fossils in America, by Jay Ellis Ransom (New York: Harper and Row, 1964). A monumental work, listing hundreds of localities, county by county, for every state. Although I must admit I have fared poorly using it in the Mid-Atlantic area, it is nevertheless a valuable resource, as it gives places to start looking and lists the formations and distinctive fossils to be found almost anywhere in the United States.

Hunting for Fossils, by Marian Murray (New York: Collier Books, 1967). Another guide to localities that hasn't produced well for me. Still, any book that gives specific information about fossils in your area is worth having.

An Illustrated Guide to Fossil Collecting 3d ed., by R. Casanova and R. Ratkevich (Happy Camp, Calif.: Naturegraph Publishers, 1981). Contains short lists of collecting localities, paleontological publications for each state, and an introduction to the hobby. Also has a good history of the science of Paleontology.

Guides to Fossils and Fossil Collecting in the Mid-Atlantic States

Eocene (Baltimore: Maryland Geologic Survey, 1901, reprinted 1963). A technical guide to the Eocene sediments and fossils of Maryland. It is most valuable to the fossil collector for its beautiful drawings of fossils. Deposits now considered to be Paleocene in age (e.g., the Aquia formation) are also covered by this volume.

Fossil Collecting in Maryland, by John D. Glaser (Baltimore: Maryland Geological Survey, 1979). Presents a small selection of well-described localities, most of which I've visited with success.

Fossil Collecting in Pennsylvania (3d ed.), by D. Hoskins, J. Inners, and J. Harper (Harrisburg: Pennsylvania Geological Survey, 1983). A superb book, brilliantly conceived and well illustrated, with detailed discussions and descriptions of many localities. Most of the ones I have visited panned out well.

Fossils of Calvert Cliffs, by W. Ashby and M. Parrish (Solomons, Md.: Calvert Marine Museum Press, 1986). A nicely illustrated popular guide to the fossils of this area, with a useful description of the different zones of fossiliferous sediments exposed in the cliffs.

Fossil Sharks of Maryland: An Illustrated Guide, by Bretton W. Kent (College Park, Md.: Bretton W. Kent, 1987). A clearly written, carefully illustrated guide to a very difficult subject—shark's tooth identification. Available at Matoaka Cottages and elsewhere in the area.

Geologic History of West Virginia, by Dudley H. Cardwell (Morgantown: West Virginia Geological Survey, Educational Series 10, 1975). A very useful guide to the geologic history, fossils, and fossiliferous formations of this state. Includes illustrations of typical genera for each period.

Guide to Common Cretaceous Fossils of Delaware, by Thomas E. Pickett (Newark: Delaware Geological Survey, Report of Investigations 21, 1972). Includes directions to localities and pictures of typical fossils from this area.

Guide to Fossil Collecting in Virginia, by Eugene K. Rader (Charlottesville: Virginia Division of Mineral Resources, 1964). Shows no localities or formations for the fossils but describes the major groups of fossils and provides a small-scale geologic map of the state.

Miocene, 2 vols.: text and fossil plates (Baltimore: Maryland Geological Survey, 1904, reprinted 1973). An exhaustive technical guide to the Miocene sediments and fossils of Maryland. Beautiful illustrations in the fossil plates volume.

Miocene Fossils of Maryland, by Harold E. Vokes (Baltimore: Maryland Geological Survey, 1957). In effect, an abridgment of the above two-volume set, with a selection of the same drawings of fossils. It might replace the two-volume set for fossil collectors if it weren't for the fact that only five kinds of sharks' teeth are shown.

Plant Fossils of West Virginia, by W. Gillespie, J. Clendening, H. Pfefferkorn (Morgantown: West Virginia Geological and Economic Survey, 1978). A very useful, profusely illustrated book. I only wish it included a list of localities.

Roadside Geology of Virginia, by Keith Frye (Missoula, Mont.: Mountain Press, 1986). Not specifically about fossil collecting, but the descriptions of local geology, presented as a series of field trips, help to put the history of local fossil-bearing rocks in perspective.

Professional Guides to Fossil Identification

Index Fossils of North America, by Hervey W. Shimer and Robert R. Shrock (Cambridge, Mass.: MIT Press, 1944). The organization of this expensive book makes it exasperating to use, and it gives little information about the organisms; but it is the most complete guide to invertebrate fossils available in one volume. Invaluable for identification.

Invertebrate Fossils, by R. Moore, C. Lalicker, and A. Fischer (New York: McGraw-Hill, 1952). A thoroughly illustrated, comprehensive guide to invertebrate fossils. It includes excellent discussions of the biology of the living organisms.

Treatise on Invertebrate Paleontology, ed. R. C. Moore (Lawrence, Kans.: Geological Society of America and the University of Kansas Press, 1954). A monumental series of many volumes, begun in 1953 and still being worked on. This highly technical work covers all of the groups of invertebrate fossils and is the ultimate source for identification. It is available in the reference section of most university libraries.

Vertebrate Paleontology and Evolution, by Robert L. Carroll (San Francisco: W. H. Freeman, 1988). Lists almost every genus of fossil vertebrate. Its 700 pages are well illustrated, but this book is not intended to help the hobbyist identify the odd tooth or bone fragment.

Miscellaneous Publications about Fossils

If You Are a Hunter of Fossils, by B. Baylor and P. Parnall (New York: Scribner's, 1980). A unique children's book that focuses on the romance of fossil hunting.

Malick's Fossils, Inc., Catalog. Current edition may be obtained from Malick's Fossils, Inc., 5514 Plymouth Road, Baltimore, Maryland 21214. Catalogs such as this one can be useful aids to identification as well as sources of publications and fossil specimens.

Rocks and Minerals Magazine. Heldref Publications, 4000 Albemarle Street, N.W., Washington, D.C. 20016. This magazine frequently publishes articles of interest to the fossil collector, including descriptions of localities.

Guides to Prehistoric Life

The Age of Birds, by Alan Feduccia (Cambridge, Mass.: Harvard University Press, 1980). A well-illustrated survey of the evolutionary history of birds. As bird fossils are comparatively rare, few guidebooks give them much mention, so this is an especially welcome book.

The Age of Mammals, by Bjorn Kurten (New York: Columbia University Press, 1971). A good account of the development of life during the Cenozoic era; not very profusely illustrated, however.

The Ecology of Fossils: An Illustrated Guide, by W. S. McKerrow (Cambridge, Mass.: MIT Press, 1978). A remarkable book, containing drawings and descriptions of over one hundred natural communities as reconstructed from fossil evidence. A British study, but many of the habitats had North American counterparts.

The Encyclopedia of Prehistoric Life, ed. R. Steel and A. Harvey (New York: McGraw-Hill, 1979). Easy to find information about Earth's history and the science of paleontology.

The Illustrated Encyclopedia of Dinosaurs, by David Norman, illus. John Sibbick (New York: Crescent Books, 1985). An excellent, magnificently illustrated book of some value to fossil collectors as it provides detailed drawings of skeletons and photographs of the fossils themselves.

Life Before Man, by Z. Burian and Z. Spinar (New York: American Heritage Press, 1972). Presents a collection of wonderful paintings by one artist. This book brings more prehistoric worlds and their inhabitants to life than any other I know.

The Macmillan Illustrated Encyclopedia of Dinosaurs and Prehistoric Animals, by Dougal Dixon et al., (New York: Macmillan, 1988). A colorfully illustrated survey of fossil vertebrates, including types that are less well known than dinosaurs.

Prehistoric World, by Richard Moody (Secaucus, N.J.: Chartwell Books, 1980). A richly produced collection of essays on a variety of topics of interest to the fossil enthusiast. One of its strengths is that it doesn't ignore marine invertebrates that lived after the appearance of land animals—a rare virtue.

The Rise of Life, by John Reader, illus. John Gurche (New York: Alfred A. Knopf, 1986). A scientifically oriented history of life and the development of the discipline of paleontology, with some inspired illustrations.

The World We Live In, by Lincoln Barnett and editorial staff (New York: Time Life, Inc., 1955). A childhood favorite, included here because of its illustrations, which will help you visualize the worlds your fossils came from.

A Sampling of Textbooks

The Earth through Time, by Harold L. Levin (Philadelphia: W. B. Saunders, 1978). A college textbook especially useful for its appendix, which gives a list of the rock formations for selected parts of the United States.

Fundamentals of Paleobotany, by Sergei V. Meyen (New York: Chapman and Hall, 1987). A comprehensive textbook on the history and development of plants.

Geological Evolution of North America (3d ed.), C. Stern, R. Carroll, and T. Clark (New York: John Wiley and Sons, 1979). This text includes a detailed account of how the rocks were formed and the continent has changed over time. Illustrations include many pictures of fossils.

Invertebrate Paleontology and Evolution (2d ed.), by E. N. K. Clarkson (London: Allen and Unwin, 1986). An outstanding paleontology text, which presents a good survey of the different groups of fossil organisms and their probable biologies. Also features an expanded section on trilobites.

Principles of Paleontology, by D. Raup and S. Stanley (San Francisco: W. H. Freeman, 1978). An outstanding college textbook, which demonstrates just how much remarkable information can be gleaned from the fossil record.

Professional Reports and Bulletins

The United States Geological Survey and state geological surveys (or equivalent organizations) publish a wide variety of material that is useful to the fossil collector. Besides geologic maps and guides to collecting, many publications provide details about local geology that can help one become acquainted with the rocks in an area and help one to locate new fossil localities. Listed below are some samples. The addresses for the institutions in the region are given in the Appendix. Contact them for free lists of the publications available.

Bedrock Geology of the Evitts Creek and Pattersons Creek Quadrangles, Maryland, Pennsylvania, and West Virginia, by W. Dewitt and G. Colton (Washington, D.C.: U.S. Geological Survey Bulletin 1173, 1964). One of the many bulletins published by and available through the U.S.G.S.; it gives priceless information about the geology and fossils of a defined area. Includes geologic maps.

Environmental History of Maryland Miocene, by R. Gernant, T. Gibson, and F. Whitmore (Baltimore: Maryland Geological Survey Guidebook 3, 1971). An in-depth study, which provides much information about the occurrence and significance of some of the fossiliferous sediments of Maryland.

Geology and Mineral Resources of Frederick County, by C. Butts and R. Edmundson (Charlottesville: Virginia Division of Mineral Resources, Bulletin 80, 1966). A professional survey of the rocks and fossils of an entire county in Virginia. It is also useful as a guide to rocks in adjacent areas. Includes a geologic map.

Geology and Mineral Resources of Southern Somerset County, Pennsylvania, by Norman K. Flint (Harrisburg: Pennsylvania Geological Survey, 1965). A description of the rock formations and their economic uses in a portion of one county. An appendix gives directions to a few localities and lists of the fossils to be found there. Includes geologic maps.

Geology of the Appalachian Valley in Virginia, 2 vols.: geologic text and fossil plates, by Charles Butts (Charlottesville: Virginia Geologic Survey, 1940, 1941). A complete guide to the rock formations of western Virginia, which extend into adjacent states. Many of the formation and fossil names have been changed, but this book is still valuable for its lists of fossil types from each formation and countless localities.

Geology of the Woodstock, Wolf Gap, Conicville, and Edinburg Quadrangles, Virginia, by R. Young and E. Rader (Charlottesville: Virginia Division of Mineral Resources, Report of Investigations 35, 1974). State geological report that includes formation and fossil information as well as geologic maps. A road log gives points of geologic interest in the area covered, including a few fossil localities.

The Lower York–James Peninsula, by Joseph Kent Roberts (Charlottesville: Virginia Geological Survey Bulletin 37, 1932). A field report that actually provides a chapter on fossil collecting in addition to its survey of the local geology.

The Onondaga Fauna of the Allegheny Region, by E. M. Kindle (Washington, D.C.: United States Geological Survey, 1912). An old technical report that discusses the fossils of one set of related rock formations. It is still useful for identification, as it includes good illustrations of specimens.

Index

Numbers in italics refer to illustrations.

AAA rule, acid solutions and, 35
Acadian orogeny, 10
Acid, fossil preparation with, 35, 64
Acrylic sprays, 34
Algae, 174. *See also* Calcareous algae; Plants
Aliquippa, Pa., 135
Allegany County, Md., 60, 61, 121–23
Alleghany orogeny, 12
Allegheny County, Pa., 138
Amber, 3
Ambridge, Pa., 135–37
Ammonites, 170
Ammonoids, 87, *89*, 90, 101, 102, 114, 169, 170. *See also* Cephalopods
Amphibians, *11*, 177
  fossils of, 11, 136, *137*, 177
Angiosperms, 173
Animals, classification of, 5–8, 24
Anthozoans, 167, 168. *See also* Corals
Ants, 50
Appalachian Mountains, 26
  formation of, 8–12, 48, 59
Aquia formation, 43, 146–49
Arrowhead, Indian, 161
Arthropods, 168. *See also* Barnacles; Crabs; Crustaceans; Eurypterids; Ostracodes; Trilobites
  molting of, 9, 98
Articulata, 168, 169
Articulations, of brachiopods, 168
Athens shale, 46
Atlantic Ocean, formation of, 12
Aves, 177

Baker, W. Va., 104–6
Barnacles, 170, 171
  fossils, 171
    Balanus concavus, *151*, 152, *153*, 154, 157, *160*, 161
Bath County, Va., 59
Beach, collecting at, 27, 28, 32
Beaver County, Pa., 135–37
Beavers, 67, 68, 69, 116
Bedding planes, 27
Beds, rock, 27

Belemnoids, 144, *145*, 169, 170. *See also* Cephalopods
Belvedere Beach, Va., 146
Berkeley County, W. Va., 117, 118
Bethany, W. Va., 142, 143
Big Bend campground, 76, 78
Big Spring, Md., 96
Birds, *17*, 177
  fossils, 147, 161, 177
Bivalvia, 173. *See also* Pelecypods
Blacksburg, Va., 46
Blastoids, 171, 172
  fossils, 124
    Pentremites sp., 124, *125*, 126
Blue-green algae, 174
Bony fish, 176, 177
Brachial valve, 169
Brachidium, 81, 168
Brachioles, crinoid, 171
Brachiopods, *9*, 168, 169
  fossils, ix, 6, 44, 53, 74, 81, 127, 129, 135, 168
    Acrospirifer murchisoni, 76, *79*, 80
    Ambocoelia nana, 84, 87, 90, 94
    Ambocoelia umbonata, 101, 102, 108, 110, 112, *113*, 114
    Anoplotheca. See *Coelospira*
    Athyris spiriferoides, 110, 112, *113*, 114, 120
    Atrypa. See *Desquamatia*
    Camarella sp., 44, 45
    Camarotoechia. See *Cupularostrum*
    Chonetinella sp., 136, *137*
    Coelospira acutiplicata, 84, 87, 90, 92, 93, 94, *95*
    Coelospira concava, 77, 78
    Composita subquadrata, 130, *131*
    Composita trinuclea, 126
    Costellirostra peculiaris, *79*, 80
    Costellirostra singularis, 81, 82, *83*
    Costispirifer arenosus, 76, *79*, *80*
    Craniops(?) sp., 60
    Cupularostrum sp., 60, *61*, 68, 80
    Cupularostrum congregatum, 119
    Cupularostrum litchfieldense, 65, 66, *71*, 72
    Cupularostrum neglectum, 62, 63

Brachiopods, fossils—*continued*
  Cyrtina sp., 96
  Cyrtina dalmani, 71, 72
  Cyrtina hamiltonensis, 106, 110
  Cyrtina varia, 75, 78, 80, 82, 83
  Dalmanella. See *Onniella*
  Delthyris sp., 106, 110, *111,* 120
  Derbyia sp., 136, *137*
  Desquamatia sp., 73
  Desquamatia reticularis, 84, 87, 90, 92, 93, 94, *95, 113,* 114, 120
  Devonochonetes sp., 114, 116
  Devonochonetes coronatus, 105, 106, 118, 119, 120
  Devonochonetes hemisphericus, 94
  Devonochonetes scitulus, 102
  Diaphragmus cestriensis, 129, 130, *131*
  Dictyoclostus (see *Reticulatia*)
  Discinisca lugubris, 151, 154, 157
  Douvillina sp., 106, 108, *109*
  Eatonia medialis, 75, 77, 78, 81, 82, *83*
  Elita fimbriata, 102, 106, 114, 120
  Elytha (see *Elita*)
  Eodevonaria arcuata, 87
  Eospirifer (see *Macropleura*)
  Hesperorthis sp., 44, 45
  Howellella cycloptera, 75, 78, 82, 83
  Howellella vanuxemi, 66
  Inflatia inflata, 124, *125,* 126
  Kozlowskiellina perlamellosa, 72, 75
  Leiorhynchus sp., 122
  Leiorhynchus limitare, 108, 118
  Leptaena "rhomboidalis," 71, 72, 75, 78, 82, 87, 90, 94
  Lindstromella aspidium, 101, 102, 114
  Lingula sp., 54, 60, 77, 78, 102, 106, 108, *113,* 114
  Linoproductus ovatus, 130, *131*
  Macropleura macropleura, 74, 75
  Mediospirifer audacula, 116, 118, 120
  Meristella sp., 71, 72, 75, 78
  Meristella lata, 80, 81, 82, *83*
  Mimella sp., 44, *45*
  Mucrospirifer sp., 122, *123*
  Mucrospirifer mucronatus, ix, 6, 101, 102, *103,* 106, 108, 110, *111,* 112, *113,* 114, 116, 117, 118, 119, 120
  Multicostella platys, 44, *45*
  Nucleospira elegans, 70, 71, 72
  Onniella sp., 54, *55*
  Onniella fertilis, 50, 52, 56, *57,* 58, 59
  Orbiculoidea(?) sp., 102
  Orbiculoidea lodiensis, 94
  Orthorhynchula linneyi, 23, 53, 54, *55*
  Orthotetes sp., *125,* 126, 129, 130, *131*
  Platyorthis planoconvexa, 75, 78, 82
  Productella sp., 122, *123*
  Productella(?) sp., 122, *123*
  Productus (see *Inflatia*)
  Pseudolingula sp., 50
  Ptychomaletoechia sp., 122, *123*
  Rafinesquina alternata, 54, 56, *57,* 58
  Rensselaeria sp., 76, 79, 80
  Rensselaeria subglobosa, 81, 82, *83*
  Reticulatia sp., 138
  Rhipidomella sp., 71, 72, 120, 130
  Rhipidomella assimilis, 82, *83*
  Rhipidomella penelope, 108, 112, *113,* 114, 118
  Rhipidomella vanuxemi, 84, *85,* 87, 93, 94, 95
  Rhynchotrema increbescens, 58
  Rhynchotreta sp., 65, 66, 71, 72
  Rhytistrophia beckii, 77, 78
  Schellwienella(?) sp., 79, 80
  Schuchertella pandora, 87, 94, 95
  Sowerbyella sp., 50
  Sowerbyella rugosa, 57, 58
  Spinatrypa sp., 106, 108, 110, *111,* 114, 120
  Spinocyrtia granulosa, 96, 97, 99, 100, 101, 102, 104, *105,* 106, 108, 110, 114, 116, 117, *118,* 120
  Spirifer sp., 126
  Spirifer increbescens, 130, *131*
  Stropheodonta (see *Protoleptostrophia*)
  Strophonella sp., 82
  Trematospira multistriata, 75
  Tropidoleptus carinatus, 99, 100, 102, *105,* 106, 108, 114, 116, 118, 119, 120, 122
  Tylothyris sp., 122, *123*
  Tylothyris mesacostalis, 122, *123*
  Uncinulus sp., 71, 72
  Zygospira sp., 54, *55,* 59
  orders of, 169
Brittle stars, 172
Brooke, Va., 146
Brooke County, W. Va., 142, 143
Brush Creek formation, 43, 135–38
Brushes, 29, 31
  brass-bristled, 34, 86, 91
Bryozoans, 44, 169
  fossils, 41, 44, 50, 54, *55, 57,* 59, 60, *61,* 62, 64, 65, *66,* 70, 72, 78, 82, *83,* 114, 118, 130, 144, 154, 157, 169
    Fenestella sp., 71, 72, 87, 94, 102, 104, 106, 108, 110, *113,* 114, 116, 117, 118, 120, 126, 129, 130, *131*
    Polypora sp., 129, 130
    Prasopora simulatrix, 56, *57,* 58, 59
    Ptylodictya sp., *105,* 106

Rhinidictya nicholsoni, 56, *57*, 58
Rhombopora sp., 138
Tretocycloecia sp., *160*, 161
identification of, 41, 169
silicified, 44, *45*
Buffalo Creek, W. Va., 142
Bullpasture Mountain, Va., 74, 75

Cacapon River, W. Va., 87, 93–95
Calcareous algae, 174
fossils, *71*, 72, 78, *79*, 80, 82, *83*
Calcareous rocks, 14
Calcite, 26, 27, 76, 86, 91, 93, 96, 104, 110, 114, 120, 138
Calcium carbonate, 10, 14, 26, 48, 76, 86, 174
Calvert Cliffs, Md., 150–54, 155
Calvert formation, 23
collecting localities, 43, 150–58
Calyx, crinoid, 171
Cambrian period, 4, 10
life of, 170, 171, 173, 174
Capon Bridge, W. Va., 93
Capon Lake, W. Va., 88–90
Carboniferous period, 4
Carcharodon carcharias, 154
Casts (internal molds), 26
Caves, 3, 67, 129, 177
Cenozoic era, 4, 5, 12, 14
divisions of, 4
life of, *17*, *19*, 167–77
Cephalon, trilobite, 176
Cephalopods, 7, 144, 169, 170
fossils, 62, 84, 86, 91, 92, 100, 101, 102, 106, 120, 135, 136, 138, 144
Agoniatites sp., 101, 102
Agoniatites(?) sp., 114
Agoniatites vanuxemi, 87, *89*, 90, 94
Belemnitella americana, 144, *145*
Michelinoceras sp., 50, 52, 54, *55*, 60, *61*, 92, 110, *113*, 114, 116
Michelinoceras subulatum, 84, *85*, 86, 87, 90
Spyroceras sp., 92, 114
Spyroceras crotalum, 107, 108
Ceratites, 170
Cetaceans, 177. *See also* Porpoises; Whales
Chalk Hill, Pa., 129
Chambered nautilus, 7, 169
Chelonia, 177
Chemung formation, 43, 122, 123
Chert, 16, 26, 44, 76, 124, 127
Chesapeake and Delaware Canal, 144
Chesapeake Bay, 10, 150–54
Chesapeake Beach, Md., 150
Chestnut Ridge, Pa., 129
Chisels, 28, *29*, 33–35

Choptank formation, 43, 150–58
Chondrichtyes, 176, 177
Chordata, 176
Cirripedia, 171. *See also* Barnacles
Classification, of animals and plants, 5–8
Cleaning fossils. *See* Fossil collecting, preparation of specimens
Clinton formation, 62
Cnidaria, 167, 170. *See also* Anthozoans; Conularids; Corals
Coal, 12, 94, 134, 142
forests, 173, 174, 177
Coeymans formation, 73
Concretions, 101, 107, 157, 171
Cones, lycopsid, 132, *133*, 134
Conglomerates, 14
Conifers, 173, 174
Conodonts, 170
Continental seas, 10, 12, 124
Conularids, 170
fossils, 170
Conularia sp., *89*, 90
Coprolites, 147, *149*, 157, 170, 175
Corallites, 167, 169
Corals, 72, 121, 167, 168, 169. *See also* Horn corals
fossils, 41, 104, 120, 130, 159
Acrocyathus sp., 124, 126
Acrocyathus floriformis, 127, *128*
Amplexus hamiltoniae, 106, 110, *111*
Astrhelia palmata, *151*, 153, 154, 157
Aulopora sp., 81, 82, *83*, 130, *131*
Balanophyllia elaborata, 147, *149*
Ceratopora sp., 105, 106
Enterolasma sp., 71, 72
Enterolasma stricta, 74, 75, 82, *83*
Favosites sp., 67, 73, 102, 121
Favosites conicus, 81, 82, *83*
Favosites helderbergiae, 68, *69*, 78
Favosites limitaris, 64, 66, 68, *69*
Favosites niagarensis, 60, *61*
Haimesastraea sp., 147, *149*
Halysites catenularia, 67, 68, *69*
Heliophyllum sp., 121
Heterophrentis sp., 92, 94, *95*, 102, 108, *113*, 114, 116, 118, 120
Lithostrotionella (see *Acrocyathus*)
Pleurodictyum sp., 87, 94, 101, 102, *103*, 106, 110, 119
Septastrea marylandica, 159, *160*, 161
Stereolasma rectum, 106
Stereostylus sp., 136, *137*, 138
Streptolasma (see *Enterolasma*)
Trachypora sp., 102, 106, 108, *109*, 120
Trachypora(?) sp., 87, 92, 94, *95*

Corals, fossils—*continued*
    Triplophyllum sp., 124, *125*, 126
    Zaphrentis sp., 94, *95*, 114
  identification of, 41
  silicified, 64, 68, *69*
Cordaitales, 174
Corriganville formation, 42, 64, 74–79
Cove Point, Md., 150
Crabs, 170, 171
  fossils, *151*, 154, 171
    Necronectes(?) sp., 157, 158
Cretaceous period, 4, 14
  collecting locality, 43, 144, 145
  life of, 174, 177
Crinoidal limestone, 73
Crinoids, 7, *9*, 171, 172
  fossils, 44, 48, *49*, 51, 52, 53, 54, *55*, 58, 68, *69*, 73, 74, 75, 76, 77, 78, 79, 80, 82, *85*, 86, 87, 90, 92, 94, *95*, 96, *97*, 100, 101, 102, 108, 110, 112, *113*, 114, 116, 118, 120, 122, *123*, 124, *125*, 126, *127*, 129, 130, *131*
    Arthroacantha sp., *99*, 100, 107, 108, *109*
    Arthroacantha punctobrachiata, 96, *97*
    Ectenocrinus simplex, *49*, 50
    Edriocrinus pocilliformis, 81, 82, *83*
    Eupachycrinus sp., 130, *131*
    Lasiocrinus(?) sp., 102
    Lasiocrinus scoparius, 96
Crocodiles, 177
  fossils, 147, *149*, 153, 158, 161, 177
    Thecachampsa sp., 147, *149*
    Thecachampsa antiqua, *151*, 154
Crown vetch, 74
Crust, Earth's, 8–12
Crustaceans, 139, 170, 171. *See also* Barnacles; Crabs; Ostracodes
Cumberland, Md., 60, 61, 122
Cyanophyta, 174
Cycads, 174
Cystoids, 171, 172
  fossils, 70, *71*, 72, 82, 83
    Pseudocrinites sp., 64, *65*, 66

Danville, Md., 121
Delaware, *40*, 43
Delaware City, Del., 144, 145
Delray, W. Va., 112
Detrick, Va., 119
Devonian period, 4, 10
  collecting localities, 42, 43, 73–123
  length of year during, 121
  life of, *9*, 167, 174, 175, 177
Diatoms, 16
Diatrymidae (extinct bird family), *17*

Dinosaurs, 12, *15*, 144, 177
  footprints of, 25, *145*
Displaying fossils, 33, 36, 37
Distortion of fossils, 44, 48, *49*, 119
Dolphins, 177

Eastover formation, 43, 155–58
Echinoderms, 171, 172. *See also* Blastoids; Crinoids; Cystoids; Echinoids; Starfish
Echinoids, 171
  fossils, 171
    Abertella aberti, 154
    Scutella, 154
Edinburg, Va., 120
Edrioasteroids, 171, 172
Effinger, Va., 44, 45
Elmer's Glue-all, *29*, 30, 34, 35, 36, 86, 135
Eocene epoch, 4, 147
  life of, *17*, 170
Epochs, geologic, 4, 5
Equipment
  fossil collecting, 27–31, *29*
  fossil preparation, 33–36
Eras, geologic, 4, 5
Erosion, 10, 14
Eurypterids, 172
Extinction, 12, 17, 67

Fairview Beach, Va., 146
Fayette County, Pa., 129–31
Fern fossils. *See* Plants
Fish (other than sharks, sharklike fish, and rays), *13*, 176, 177
  fossils, 25, 34, 116, 139, 141, 154, 157, 158, 161
    Diodon sp., 161
    Paralbula sp., 147, *149*
    Phyllodus toliapicus, 147, *149*
    Pogonias sp., *151*, 154
    swordfish, 157
Fissile shale, 27, 91, 93, 96, 135
Flag Ponds Nature Park, Md., 150, 153
Flintstone, Md., 122, 123
Fort Frederick State Park, Md., 96, 97
Fossil collecting, 23–37
  display of specimens, 36, 37
  equipment, 27–31, *29*
  ethics of, 31, 32
  hazards of, 31, 33, 152, 157
  labeling of specimens, 37
  preparation of specimens, 33–36, 86
  storage of specimens, 36, 37
Fossils, ix, x
  age of, 3–5
  collecting, 23–37

defined, 3
distortion of, 44, 48, *49*, 119
guide, 25, 66, 76, 80
identification of, 24, 25, 41
index (*see* Fossils, guide)
names of, 5–8
number of species known, 18
preservation of, 3, 14–18
pyritized, 84, 86, 101, 107
silicified, 26, 35, 44, 69, 74, 81
uses of, x
Franklin, W. Va., 53, 70, 91, 92
Frederick County, Va., 98–100, 115, 116
Fredericksburg, Va., 146, 155
Fulks Run, Va., 73

Gainesboro, Va., 115, 116
Gap Mills, W. Va., 124–26
Gastropods, 153, 172, 175
    fossils, 44, 58, 74, *125*, 126, 135, 144, *145*, 155
        Bembexia sp., *113*, 114, 118, 120
        Bembexia ella, *99*, 100, 102, 116
        Bucanella sp., 102
        Bulimorpha sp., 136, *137*
        Crenistriella sp., 102
        Crepidula sp., *160*, 161
        Cyclonema sp., 122, *123*
        Cymatospira sp., 136, *137*
        Diodora sp., *160*, 161
        Ecphora gardnerae, *151*, 153, 154, 158, 159, *160*, 161
        Epitonium virginianum, 147, *149*
        Holopea sp., 122
        Hormotoma (see *Murchisonia*)
        Liospira micula, 59
        Loxonema sp., *85*, 87, 90, 92
        Loxonema delphicola, 100, 112, *113*, 114, 116
        Loxonema hamiltoniae, 102, 106, 110, 114
        Lunatia (see *Natica*)
        Meekospira sp., 136, 138
        Murchisonia sp., 122, *123*
        Murchisonia marylandica, 60, *61*
        Natica marylandica, 147, *149*
        Platyceras sp., *71*, 72, 75, 76, 77, 78, *79*, 80, 87, 94, *99*, 100, 108, *113*, 114, 122
        Platyostoma sp., 87, 94
        Pleurotomaria (see *Bembexia*)
        Scala (see *Epitonium*)
        Shansiella sp., 136, *137*, 138
        Sinuites sp., 54
        Strobeus sp., 130, 136, *137*
        Turritella sp., *149*
        Turritella humerosa, 147

Turritella mortoni, 147, *149*
Turritella plebeia, *151*, 154, 158, 161
pteropods, 175
Geodes, 81
Geological surveys, 23, 24
    addresses of, 181
    publications of, 23, 25, 188, 189
Geologic maps, 23–25
Geologic time, 4, 5, 136
    how measured, 5
    scale of, 4
Germany Valley Overlook, W. Va., 53–55
Ginkgoes, 174
Glue, *29*, 30, 34, 35, 134
Goggles, safety, 28, *29*
Goniatites, 170
Gore, Va., 98–100
Governor's Run, Md., 152
Graptolites, 172
    fossils, 46, 172
        Climacograptus sp., 46, 50, 54, *55*
Great white shark, 154
Green algae, 174
Greenbrier group, 43, 124–28
Greenbrier limestone, 129
Growth lines, horn corals, 121
Guide fossils, 25, 66, 76, 80
Gymnosperms, 173–75

Hamilton formation, 115
Hammers, 28, *29*
Hampshire County, W. Va., 64–66, 88–90, 93–95, 112–14
Hand lenses, *29*, 30
Hardening of soft rocks, 36, 86
Hardy County, W. Va., 48–50, 67–69, 84–87, 101–11
Hazards of fossil collecting, 31–33, 152, 157
HCl (muriatic acid), 35
Hedgesville, W. Va., 117, 118
Helderberg group, 23, 64, 73, 74, 76, 78
    collecting locality, 42, 64, 65
Hemichordata, 172
Hexacorals, 167
Highland County, Va., 74, 75, 81–83
Hillsboro, W. Va., 127, 128
Holocene epoch, 4
Holothuroidea, 172
Horn corals, 7, *9*, 121, 167, 168
    fossils, *71*, 72, 74, 75, 82, *83*, 84, *85*, 87, 90, 92, 94, *95*, 102, 106, 108, *111*, *113*, 114, 116, 118, 120, 121, 124, *125*, 127, *128*, 130, *131*, 135, 136, *137*
    growth lines on, 121

Horse, tooth of, 157, 158
Horseshoe crabs, 176
Horsetails, fossil. *See* Plants
Horsetails, living, 135
Hyolithids, 173
    fossils, 173
        Hyolithes sp., 94, *95*

Iapetus Ocean, 10
Ice Age, 3, 14
Igneous rocks, 12, 23
Inarticulata, 168, 169
Index fossils. *See* Guide fossils
Indian Field Creek, Va., 159–63
Insects, 168
    fossils of, 3
Iron ore, 62
Iron oxide residues on fossils, 46, 48, 51, 96, 122
Isaacs Creek, Va., 115, 116

Jackson River, Va., 81
Juniata formation, 59
Jurassic period, 4
    life of, *13, 15,* 171, 177

Kanawha formation, 43, 132–34
Keenan, W. Va., 124–26
Keyser formation, 64, 74
    collecting localities, 42, 64–72, 74
King George, Va., 155
King George County, Va., 146–49

Labeling of specimens, 37
Labyrinthodonts, *11,* 136, *137*
Lebanon County, Pa., 51, 52
Lenoir limestone, 44
Lexington, Va., 44, 56–58
Liberty Hall formation, 42, 46, 47
Lickdale, Pa., 51
Licking Creek formation, 42, 64, 81–83
Limestone, 23, 26, 86
    crinoidal, 73
    decalcified, 84, 86
    formation of, 10, 14, 16
    fossils in, 26, 35
Limulus, 176
Lincolnshire formation, 23
    collecting locality, 42, 44, 45
Lingulida, 169
Locust Creek, W. Va., 127, 128
Lophophores, brachiopod, 81, 169
Lost City, W. Va., 107–9, 112
Lost River, W. Va., 67–69, 84–87

Lusters Gate, Va., 46, 47
Lycopsids. *See* Plants

McCoy's Ferry, Md., 96
McKenzie formation, 42, 60, 61
Mahantango formation, ix, 27
    collecting localities, 43, 96, 98–121
Mahoning formation, 43, 135–37
Malacostraca, 171
Mammals, fossils of, 161, 177. *See also* Horses; Porpoises; Whales
Mantle, Earth's, 8
Maps, 24
    geologic, 23–25
    localities in book, *40*
    road, 30, 31
Marcellus formation, 86, 93
    collecting locality, 43, 96, 97
Marlborough Point, Va., 146, 147
Marlinton, W. Va., 132–34
Martinsburg, W. Va., 117
Martinsburg formation, 23, 53
    collecting localities, 42, 48–52
Maryland, *40,* 42, 43
Massanutten Mountain, Va., 119
Matoaka Cottages, Md., 150, 152
Matrix, 26
    removal of excess, 33–36
Mauch Chunk formation, 23
    collecting locality, 43, 129, 131
Members of rock formations, 23
Merostomata, 172
Mesozoic era, 4, 12–15
    divisions of, 4
    life of, *13, 15,* 167–77
Metamorphic rocks, 24
Miocene epoch, 4
    collecting localities, 43, 150–58
    life of, *19*
Mississippian period, 4
    collecting localities, 43, 124–31
    life of, 170
Molds, 26
Mollusks, 169, 173. *See also* Cephalopods; Gastropods; Pelecypods, Pteropods
Molting, of trilobites, *9,* 98
Monongahela group, 43, 139–43
Monroe County, W. Va., 62, 63, 124–26
Monterey, Va., 81–83
Montgomery County, Va., 46, 47
Moorefield, W. Va., 110, 111
Mountains, formation of, 8
Mount Laurel formation, 144, 145
Mud cracks, 26

Muriatic acid, 35, 64
Muscle scars, of pelecypod, 173

Nautiloids, 7, 50, 52, 54, *55*, 60, *61*, 62, 84, *85*, 86, 87, 90, 91, 92, 100, 101, 106, 107, 108, 110, *113*, 114, 116, 120, 135, 136, 138, 169, 170 (*see also* Cephalopods)
Needmore formation, 23
    collecting localities, 42, 43, 84–95
New Castle County, Del., 144, 145
New Creek formation, 42, 64, 73, 74
New Jersey, 17
New Scotland formation, 74
New York State, 88, 172
North Fork Mountain, W. Va., 53
North River, W. Va., 112

Ohio County, W. Va., 139–41
Ohio River, 135–38
Oligocene epoch, 4
Onondaga formation, 88
Ordovician period, 4, 10, 59
    collecting localities, 42, 44–59
    life of, 7, 167, 169, 170, 171, 172, 176
Oriskany formation, 23
    collecting localities, 42, 76–80, 86
Orogeny, defined, 10
Orthida, 169
Orthorhynchula zone (Martinsburg and Reedsville formations), 23, 53
Osteichthyes, 176, 177
Ostracodes, 170, 171
    fossils, 30, 54, 88, 170, 171
        Bollia sp., 90
        Kloedenella sp., 60, *61*
        Kloedenia(?) sp., 90
        Leperditia sp., 64, *65*, 66
        Zygobolbina conradi, 62, *63*
Oysters, 147, 168, 173. *See also* Pelecypods

Paleocene epoch, 4
    collecting locality, 43, 146–49
Paleontology, x, 41
Paleozoic era, 4, 10–12
    divisions of, 4
    life of, 7, *9*, *11*, 167–77
    rocks from, 26
Pangaea, 12
Pedicles, of brachiopod, 168
Pedicle valve, 168
Pelecypods, 6, 153, 168, 173
    fossils, 62, *99*, 100, 101, 104, 135, 144, *145*, 148, 155, 159, 173
        Actinopteria decussata, 114, 118, 120
        Allorisma terminale, 138

Ambonychia sp., *49*, 50, 53, 54, *55*
Anadara sp., 158, 161
Anadara staminea, 154
Anomia tellinoides, 144, *145*
Astarte sp., 158
Aviculopecten cancellatus, 122
Byssonychia (see *Ambonychia*)
Chesapecten jeffersonius, 159, 161
Chesapecten nefrens, *151*, 152, 154, 158, 159, *160*, 161
Cleidophorus nitidus, 60, *61*
Corbula sp., *151*, 154
Cornellites flabella, 102, *103*, 116, 118, 120
Crassatella sp., 154, 161
Crassatella alaeformis, 148, *149*
Cucullaea gigantea, 148
Cypricardella sp., 104, *105*, 106, 108, 114
Dosiniopsis lenticularis, 148
Exogyra costata, 144, *145*
Glycymeris sp., 148, 154, 161
Grammysia bisulcata, 102, 116
Grammysiodea alveata, 102, 114
Gryphaea (see *Pycnodonte*)
Isognomon maxillata, 154, *156*, 157, 158
Kuphus(?) sp., *160*, 161
Lirophora sp., 158
Lyriopecten sp., 90
Lyriopecten orbiculatus, 108
Mercenaria sp., 154, 158, 161
Modiomorpha sp., 87, 100, 104, 106
Modiomorpha concentrica, 101, 102, 110, 114, 118
Nuculites oblongatus, 101, 102, 116
Nuculoidea sp., 90, 102, 106, 114
Nuculopsis sp., 136, *137*, 138
Orthonota sp., 102
Orthonota undulata, *113*, 114, 116
Ostrea sp., *145*, 161
Ostrea alepidota, 148
Ostrea compressirostra, 148
Ostrea mesenterica, 144, *145*
Ostrea percrassa, 154
Ostrea vomer, 144, *145*
Palaeoneilo sp., 101, 102
Palaeoneilo emarginata, 106
Panenka (see *Praecardium*)
Phestia sp., 136, *137*
Phestia(?) sp., *113*, 114
Placopecten clintonius, *160*, 161
Praecardium multiradiatum, 84, *85*, 87, 88, 90, 92, 94
Pterinea sp., 116
Ptychopteria sp., 104, *105*, 106
Pycnodonte sp., 144, *145*

Pelecypods, fossils—*continued*
    Venericardia sp., 161
    Venericardia planicosta, 148
    Wilkingia, 130, *131*
Pendleton County, W. Va., 53–55, 70–72, 76–80, 91, 92
Pennsylvania, 10, *40,* 42, 43
Pennsylvanian period, 4
    collecting localities, 43, 132–43
    life of, *11,* 134, 167, 173, 174, 176, 177
Periods, geologic, 4, 5
Permian period, 4
    life of, 167, 177
Perry, W. Va., 48–50
Petersburg, W. Va., 76
Petrified wood, 148, 154, 157, 158, 173
Photography, of fossils, 30, 32
Phyllicopsida, 173, 175
Pinnules, crinoid, 171
Plants, 5, 6, 173–75
    algae, calcareous, *71, 72, 78, 79,* 80, 82, *83*
    classification of, 5–8, 24, 174
    ferns
        seed, *11,* 132, *133,* 134, 136, *137,* 173
        true or herbaceous, 136, *137,* 173
    fossils, 34, *71, 72,* 94, 96, 97, 122, *123,* 132, 142, 148, 173–75
        Alethopteris(?) sp., *140,* 141
        Alethopteris lonchitica, 133, 134
        Alethopteris zeilleri, 140, 141
        Alloiopteris sp., *140,* 141
        Annularia sp., *140,* 141, 170
        Annularia stellata, 142, *143*
        Calamites sp., 132, *133,* 134, 135, 136, *137,* 141, 174
        Cordaites sp., 141, 173
        Lepidodendron sp., 132, *133,* 134, 174
        Lepidophyllum sp., *140,* 141, 174
        Lepidostrobus sp., 132, *133,* 134, 174
        Mariopteris sp., 132, *133,* 134
        Neuropteris sp., 132, *133,* 134, 135, 136, *137,* 142, *143*
        Neuropteris scheuchzeri, 142, *143*
        Pecopteris sp., 135, 136, *137*
        Pecopteris aborescens, 139, *140,* 141
        Stigmaria sp., 134, 174
        Stigmaria ficoides, 132, *133,* 134
        Trigonocarpus sp., *133,* 134
    lycopsids, or scale trees, 132, *133,* 134, 173–75
    petrified wood, 148, 154, 157, 158, 173
    pollen fossils, 173
    receptaculitids, 174
    seed fossils, *133,* 134, 148

sphenopsids, or horsetails, 132, *133,* 134, 135, 136, *137,* 173–75
stromatolites, 174
Plate tectonics, 8–12, 14
Pleistocene epoch, 4. *See also* Ice Age
Plesiosaur, *13*
Plications, brachiopod, 168
Pliocene epoch, 4
    collecting locality, 43, 159–63
Pocahontas County, W. Va., 127, 128, 132–34
Polar ice caps, 14
Polishing, of rock surfaces, 91, 114, 127, 128
Pollen, fossil, 173
Polyps, coral, 167
Poriferans, 175
    fossils, *71, 72,* 175
Porpoises, fossils, *151,* 153, 154, 158, 177
Precambrian eras, 4
    life of, 173, 174, 175
    rocks, 23
Preparation, of fossils, 33–36, 86
Pteridospermales, 173, 175
Pteropods, 175
    fossils, 87, 118
Pterosaurs, fossils, 144
Pygidium, trilobite, 176
Pyrite (fool's gold), 84, 91
Pyritized fossils, 84, 86, 91, 101, 107

Quartz, 16, 127. *See also* Silica
    crystals of, 81, 110, 115
Quaternary period, 4, 14
    divisions of, 4

Radiolarians, 16
Radiometric dating, 5
Ray fin fish, 13
Rays, 176
    fossils, 148, 158, 161
        Aetobatis sp., *151,* 154, 158
        Myliobatis sp., 148, *149, 151,* 154, 158
Receptaculitids, 174
Red beds, 25, 59
Reedsville formation, 42, 53–59
Reefs, coral, 167
    stromatoporoid, 175
Reptiles, 177. See also Crocodiles; Dinosaurs; Plesiosaur; Pterosaurs; Turtles
Rhynchonellida, 169
Ridgeley sandstone, 76, 80
Rifts, crustal, 12, 25
Ring-necked snakes, 50
Rio, W. Va., 112–14
Ripple marks, 26
Riprap, 159, *160,* 161

Rochester formation, 42, 60, 61
Rock bags, *29*, 30
Rockbridge County, Va., 44, 45, 56–58
Rock hammers. *See* Hammers
Rockingham County, Va., 73
Rock saws. *See* Saws
Rocky Gap Park, Md., 60
Romney, W. Va., 64–66
Rose Hill formation, 42, 62, 63
Rugose corals, 167. *See also* Horn corals

St. Leonard, Md., 152
St. Mary's formation, 43, 150–58
Sand dollars, 154, 171
Sandstone, 23, 25–27
 formation of, 10, 14, 16
 kinds of, 16
Sauropod dinosaurs, *15*
Saws, 34, 35, 91, 114, 127
Scales, fish, 25, 157, 158, 170
Scale trees. *See* Plants
Scallops, 173. *See also* Pelecypods
 fossils, *151,* 152, 159, *160,* 161
Scientists Cliffs, Md., 152
Scleractinian corals, 167, 168
Scree, defined, 25
Sea butterflies, 175
Sea cucumbers, 172
Seafloor spreading, 12
Sea lilies, 171. *See also* Crinoids
Sea scorpions, 172
Sediment, 5, 10, 12, 14, 16
Sedimentary rocks, 5, 23, 25–27
 formation of, 14, 16
Seed ferns. *See* Plants
Seeds, fossil, *133,* 134, 148
Selachii, 177
Septum (plural, septa)
 cephalopod, 169, 176
 coral, 72, 167
Sewickley, Pa., 138
Shale, 23, 25–27
 fissile, 27, 91, 93, 96, 135
 formation of, 10, 14, 16
Sharklike fish, 135
 fossils, 135, 136, 137
Sharks, *19,* 153
 fossils, 148, *151,* 153, 154, 159, 161
  Alopias grandis, 154, 158
  Carcharhinus egertoni, 154, 158, *160,* 161, 162
  Carcharias (see *Carcharhinus*)
  Carcharodon sp., *19,* 27, 176
  Carcharodon megalodon, 135, *151,* 154, 157, 158

 Eugomphodus sp., 154, 158, *160,* 162
 Galeocerdo aduncus, 154, 158, *160,* 161, 162
 Galeocerdo contortus, 151, 154, 158
 Hemipristis serra, 151, 154, 158, 162
 Isurus crassus, 154, 158
 Isurus desori, 151, 154, 158, 159, 162
 Isurus hastalis, 154, 158, 159, *160,* 162
 Lamna obliqua, 148 (see also *Otodus obliquus*)
 Negaprion sp., 154, 158
 Notidanus (see *Notorhynchus*)
 Notorhynchus primigenius, 154, 158, *160,* 162
 Odontaspis sp., 148, *149*
 Otodus obliquus, 147, 148, *149*
 Oxyrhina (see *Isurus*)
 Scapanorhynchus elegans, 147, 148, *149*
 great white, 154
Sharks' teeth, collecting, 27, 28, 30, 147
Shenandoah County, Va., 119, 120
Shenandoah Mountain, Va., 74
Shenandoah Valley, 10, 26
Sieves, 30
Silica, 14, 16, 26
Silicified fossils, 26, 35, 44, 64, 74, 81
Siltstone, 27
Silurian period, 4, 10
 collecting localities, 42, 60–72
 life of, *7,* 67, 167, 171, 172, 175
Siphuncle, cephalopod, 60, 169, 170
Skates, 176
Smoke Hole, W. Va., 76–80
Snails, 172, 173. *See also* Gastropods
Species, animal and plant, 5, 6, 24, 41
 longevity of, 18
 numbers of, 18
Sphenopsids. *See* Plants
Spicules, sponge, 175
Spiriferid brachiopods, 6, 59, 117, 169
Sponges, 16, 175. *See also* Poriferans
Stafford County, Va., 146–49
Starfish, fossils of, 51, 172
Steinkerns, 26, 118
Stelleroidea, 172
Stigmaria, 133, 134, 174
Storage of fossils, 36, 37
Strasburg, Va., 119
Stratford Hall, Va., 155, 157
Stromatolites, 174
Stromatoporoids, 175
 fossils, 68, *71,* 72, 81, 82, *83*
Strophomenida, 169
Swatara Gap, Pa., 51, 52
Swordfish fossil, 157

Tabulate corals, 167, 168
Taconic orogeny, 10
Tectonics, plate, 8–12, 14
Teeth, fossil, 27, 28, 30, 137
   crocodile, 147, *149, 151,* 153, 154
   fish (other than sharks and rays), 25, 147, *149*
   horse, 157, 158, 177
   mastodon, 177
   porpoise, *151,* 154, 158, 177
   ray, 148, *149, 151,* 154, 158, 176
   sharklike fish, 135, 136, *137*
   shark, 27, 28, 100, 135, 144, 147, 148, *149, 151,* 153, 154, 159, *160,* 161, 162, 176
   whale, 153, 154, 158, *160,* 162, 177
Tentaculitids, 175
   fossils, 76, 119, 175
      Tentaculites sp., 62, *79,* 80, 119
Terebratulida, 169
Tertiary period, 4
   collecting localities, 43, 146–63
   divisions of, 4
   life of, *17, 19,* 167, 170, 176, 177
Tetracorals, 167
Theropod dinosaurs, *15*
Thorn Creek, W. Va., 70–72
Tides and fossil collecting, 27, 157, 161
Tonoloway formation, 42, 64, 65
Tools
   fossil collecting, 28–31, *29*
   fossil preparation, 33–36
Trace fossils, 25, 94, 175
Tracheophyta, 173, 174
Tree oysters, 154, *156,* 157, 158
Triassic period, 4, 12
   fossils, 25, 174, 176
   life of, 167, 177
   red beds from, 25
Trilobites, *7, 9,* 24, 176
   fossils, 44, 48, 51, 53, 72, 84, 98, 115, 176
      Acidaspis (see *Odontopleura*)
      Ampyxina scarabeus, 46, 47
      Basidechenella rowi, 87, *89,* 90
      Calliops sp., 50, 54, *57,* 58
      Calymene sp., 60, *61*
      Calymene cresapensis, 62, *63*
      Ceraurus sp., 54, *57,* 58
      Coronura aspectans, 87, 93, 94, *95*
      Cryptolithus sp., 56, 58
      Cryptolithus tesselatus, 48, *49,* 50, 51, *52*
      Dalmanites sp., 62, 80, 82, 84, *85,* 87, *89,* 90
      Dalmanites pleuroptyx, 77, 78
      Dechenella sp., 106, 108
      Dionide holdeni, 46, 47

      Dipleura (see *Trimerus*)
      Flexicalymene sp., 52
      Flexicalymene granulosa, 48, *49,* 50, 53, 54, 55
      Greenops boothi, 102, 104, 106, 107, 108, 112, *113,* 114, 118, 119
      Homotelus sp., 44
      Isotelus sp., 50, 53, 54, 58
      Kaskia sp., 129, 130, *131*
      Kaskia(?) sp., *125,* 126
      Lichas (similar to), *89,* 90
      Liocalymene clintoni, 62, *63*
      Odontocephalus aegeria, 84, *85,* 87, 90, 93, 94, *95*
      Odontopleura sp., *49,* 50
      Odontopleura callicera, 84, *85, 87,* 90
      Phacops cristata, 84, *85,* 87, 90, 91, 92
      Phacops logani, 77, 78
      Phacops rana, 24, 84, *85,* 86, 87, 90, 91, 92, 93, 94, *95,* 98, *99,* 100, 102, 104, 106, 107, 108, 110, 112, *113,* 114, 115, 116, 117, 118, 120, 122
      Proetus (see *Basidechenella*)
      Triarthus sp., 50
      Trimerus sp., 62, *63*
      Trimerus dekayi, 102, 104, 106, 107, 108, 110, *111,* 112, *113,* 114, 115, 116, 120
      Trimerus vanuxemi, 76, 77, 78, *79,* 80
   molting of, 9, 98
Turtles
   fossils, 148, 158, *160,* 167, 177
      Asperidites virginiana, 148, *149*
      Trionyx (see *Asperidites*)

Uniontown, Pa., 129–31

Vertebrates, 176, 177. *See also* Birds; Crocodiles; Dinosaurs; Fish; Mammals; Plesiosaur; Porpoises; Pterosaurs; Sharklike fish; Sharks; Turtles; Whales
Virginia, *40,* 42, 43
Volcanoes, 8, 10

Waders, hip or chest, 27, 161
Waiteville, W. Va., 62, 63
Wardensville, W. Va., 67–69, 84, 88, 101–3, 112
Warm Springs, Va., 59
Washington County, Md., 96, 97
Waterlick, Va., 119
Weather conditions, fossil collecting and, 27, 32, 33
Westmoreland County, Va., 155–58
Westmoreland State Park, Va., 155, 157

West Virginia, *40,* 42, 43
Whales, *19,* 177
  ear bones, *163*
  fossils, 27, 153, 154, 158, 159, *162, 163,* 177
    Cetotherium, 19
    Orycterocetus crocodilinus, 160, 162
Wheeling, W. Va., 139–41
Williamsburg, Va., 159
Williams River, W. Va., 132
Winchester, Va., 98, 115, 116

Wymps Gap limestone member (Mauch Chunk formation), 23, 129

Yellow Spring, W. Va., 93
York River, Va., 159–63
Yorktown, Va., 159
Yorktown formation, 43, 159–63

Zones, in rock formations, 23
Zooecium (plural zooecia), bryozoan, 129, 169

Index   **201**

Fossil Collecting in the Mid-Atlantic States

Designed by Ann Walston

Composed by Brushwood Graphics, Inc.,
in Meridien text and display

Printed by The Maple Press Company
on 80-lb. Paloma Matte

Bound by American Trade Bindery